Springer Tracts in Modern Physics
Volume 148

Managing Editor: G. Höhler, Karlsruhe

Editors: J. Kühn, Karlsruhe
Th. Müller, Karlsruhe
R. D. Peccei, Los Angeles
F. Steiner, Ulm
J. Trümper, Garching
P. Wölfle, Karlsruhe

Honorary Editor: E. A. Niekisch, Jülich

Springer
Berlin
Heidelberg
New York
Barcelona
Hong Kong
London
Milan
Paris
Singapore
Tokyo

Springer Tracts in Modern Physics

Springer Tracts in Modern Physics provides comprehensive and critical reviews of topics of current interest in physics. The following fields are emphasized: elementary particle physics, solid-state physics, complex systems, and fundamental astrophysics.
Suitable reviews of other fields can also be accepted. The editors encourage prospective authors to correspond with them in advance of submitting an article. For reviews of topics belonging to the above mentioned fields, they should address the responsible editor, otherwise the managing editor.
See also http://www.springer.de/phys/books/stmp.html

Managing Editor

Gerhard Höhler

Institut für Theoretische Teilchenphysik
Universität Karlsruhe
Postfach 69 80
D-76128 Karlsruhe, Germany
Phone: +49 (7 21) 6 08 33 75
Fax: +49 (7 21) 37 07 26
Email: gerhard.hoehler@physik.uni-karlsruhe.de
http://www-ttp.physik.uni-karlsruhe.de/

Elementary Particle Physics, Editors

Johann H. Kühn

Institut für Theoretische Teilchenphysik
Universität Karlsruhe
Postfach 69 80
D-76128 Karlsruhe, Germany
Phone: +49 (7 21) 6 08 33 72
Fax: +49 (7 21) 37 07 26
Email: johann.kuehn@physik.uni-karlsruhe.de
http://www-ttp.physik.uni-karlsruhe.de/~jk

Thomas Müller

Institut für Experimentelle Kernphysik
Fakultät für Physik
Universität Karlsruhe
Postfach 69 80
D-76128 Karlsruhe, Germany
Phone: +49 (7 21) 6 08 35 24
Fax: +49 (7 21) 6 07 26 21
Email: thomas.muller@physik.uni-karlsruhe.de
http://www-ekp.physik.uni-karlsruhe.de

Roberto Peccei

Department of Physics
University of California, Los Angeles
405 Hilgard Avenue
Los Angeles, CA 90024-1547, USA
Phone: +1 310 825 1042
Fax: +1 310 825 9368
Email: peccei@physics.ucla.edu
http://www.physics.ucla.edu/faculty/ladder/peccei.html

Solid-State Physics, Editor

Peter Wölfle

Institut für Theorie der Kondensierten Materie
Universität Karlsruhe
Postfach 69 80
D-76128 Karlsruhe, Germany
Phone: +49 (7 21) 6 08 35 90
Fax: +49 (7 21) 69 81 50
Email: woelfle@tkm.physik.uni-karlsruhe.de
http://www-tkm.physik.uni-karlsruhe.de

Complex Systems, Editor

Frank Steiner

Abteilung Theoretische Physik
Universität Ulm
Albert-Einstein-Allee 11
D-89069 Ulm, Germany
Phone: +49 (7 31) 5 02 29 10
Fax: +49 (7 31) 5 02 29 24
Email: steiner@physik.uni-ulm.de
http://www.physik.uni-ulm.de/theo/theophys.html

Fundamental Astrophysics, Editor

Joachim Trümper

Max-Planck-Institut für Extraterrestrische Physik
Postfach 16 03
D-85740 Garching, Germany
Phone: +49 (89) 32 99 35 59
Fax: +49 (89) 32 99 35 69
Email: jtrumper@mpe-garching.mpg.de
http://www.mpe-garching.mpg.de/index.html

Metin Tolan

X-Ray Scattering from Soft-Matter Thin Films

Materials Science and Basic Research

With 98 Figures

Springer

Dr. Metin Tolan
Universität Kiel
Institut für Experimentelle und
Angewandte Physik
Leibnizstrasse 19
D-24098 Kiel
Email: pex40@rz.uni-kiel.de

Physics and Astronomy Classification Scheme (PACS):
61.10.-i, 61.41.+e, 68.55.Jk

ISSN 0081-3869
ISBN 3-540-65182-9 Springer-Verlag Berlin Heidelberg New York

Library of Congress Cataloging-in-Publication Data applied for.

Die Deutsche Bibliothek – CIP Einheitsaufnahme

Tolan, Metin: X-ray scattering from soft matter thin films: materials science and basic research/Metin Tolan. – Berlin; Heidelberg; New York; Barcelona; Hong Kong; London; Milan; Paris; Singapore; Tokyo: Springer, 1999
(Springer tracts in modern physics; Vol. 148)
ISBN 3-540-65182-9

Typesetting: Camera-ready copy by the author using a Springer TEX macro package
Cover design: *design & production* GmbH, Heidelberg

SPIN: 10688533 56/3144 - 5 4 3 2 1 0 – Printed on acid-free paper

Preface

Grazing-angle x-ray scattering and the physics of soft-matter films in general are fields of modern condensed-matter research which have grown rapidly in the past decades. This book is intended to give a review of the combination of the two. The state of the art in grazing-angle x-ray scattering from soft-matter thin films is reported.

At the beginning, an introduction to the formalism of x-ray reflectivity is given. Here my intention was to present a strict description for the case of rough interfaces because quite different schemes may be found in the literature. Other theoretical chapters of this book deal with new inversion techniques for x-ray reflectivity data, and off-specular and coherent x-ray scattering. Whereas the theory of off-specular scattering is outlined briefly, the description of x-ray scattering with coherent radiation is given in some more detail since it is a very young and promising research field, particularly for the investigation of soft-matter films.

The selection of the examples was mainly driven by the fact that x-ray scattering is almost the only possible way to obtain structural information on atomic length scales from these soft-matter films. Moreover, while the atomic dimensions are sampled, mesoscopic lengths up to several microns are accessible, too, owing to the high resolution in the regime of glancing angles. One of the aims of this book is to highlight this nice fact of x-ray scattering which makes this probe so unique and powerful. The underlying physics of all examples is quite different, showing the various research areas where x-ray scattering has been successfully applied to shed light on new phenomena.

All examples are discussed in great detail. Thus, many parts of this book may also be used for tutorial courses or lectures. This purpose is supported by many cross-references, particularly between the chapters dealing with experiments and theory, and a large bibliography, where I apologize for each piece of work which should have been quoted but which I accidently have overlooked.

Acknowledgments

I am indebted to Werner Press for continuous support during the last five years and the many possibilities that he has opened for me. I profited enormously from the power (!!) of Olaf Bahr, Alfons Doerr, Jens-Peter Schlomka, and – last but not least – Oliver Hermann Seeck, who supported my work whenever possible. Further, I have to thank many other former and present members of the "Press-Gang", namely (in alphabetical order) Bernd Asmussen, Detlef Bahr, Dirk Balszunat, Rainer Bloch, Lutz Brügemann, Bernd Burandt, Wolfgang Caliebe, Eberhard Findeisen, Stefan Grieger, Christian Gutt, Dirk Hupfeld, Christoph Jensen, Inger Kaendler, Markus Kalning, Wolfgang Kison, Gregor König, Hauke Krull, Michael Lütt, Martin Müller, Volker Nitz, Christian Nöldecke, Ralf Paproth, Lutz Schwalowsky, Tilo Seydel, Michael Sprung, Rainer Stabenow, Jochim Stettner, Matthias Strzelczyk, Jörg Süßenbach, and Uwe Zimmermann, for their pleasant collaboration, for their help with numerous computer problems, and for their help concerning all the other problems that may happen in the daily life.

My special thanks go to Sunil Sinha (Advanced Photon Source), Miriam Rafailovich, and Jonathan Sokolov (SUNY Stony Brook), who kindly supported all of my longer visits to their research labs. I have to thank the Deutsche Forschungsgemeinschaft, the former BMFT, Argonne National Laboratories, Exxon Research Laboratories, Brookhaven National Laboratories, and the State University of New York at Stony Brook for very generous financial support for all of my visits. I am grateful to Doon Gibbs and John Axe (Brookhaven National Laboratories) for their hospitality. It was further a great pleasure to collaborate with Alain Gibaud, Zhixin Li, Diep Nygun, Giacomo Vacca, Shichun Qu, Jin Wang, Apollo Wong, Xiao Zhong Wu, and Weizhong Zhao during my stays in the US. Special thanks also to Heribert Lorenz and Jörg Peter Kotthaus (LMU Munich) for their continuous support with high-quality grating samples.

Finally, I would like to thank Werner Press, Jens-Peter Schlomka, and Ricarda Opitz (Humboldt University, Berlin; now at AMOLF, Amsterdam) for a (certainly very time-consuming!) critical reading of the manuscript.

Kiel, August 1998 *Metin Tolan*

Contents

List of Acronyms

AFM	atomic force microscopy
APS	Advanced Photon Source
BS	Beckmann–Spizzichino
CdA	cadmium arachidate
DWBA	distorted-wave Born approximation
ESRF	European Synchrotron Radiation Facility
FWHM	full-width half-maximum
GID	grazing-incidence diffraction
HASYLAB	Hamburger Synchrotronstrahlungslabor
KPZ	Kardar–Parisi–Zhang
LB	Langmuir–Blodgett
MBE	molecular beam epitaxy
MCF	mutual coherence function
MOVPE	metalorganic vapor phase epitaxy
NC	Névot–Croce
NSLS	National Synchrotron Light Source
PEP	polyethylene–propylene
PMMA	polymethylmethacrylate
PS	polystyrene
PSD	power spectral density
PVP	polyvinylpyridine
RHEED	reflection high-energy electron diffraction
rms	root-mean-square
STM	scanning tunneling microscopy
TEM	transmission electron microscopy
UHP	upper half-plane
vdW	van der Waals
XPCS	x-ray photon correlation spectroscopy

1. Introduction

The investigation of surfaces and interfaces with x-ray scattering methods is a field that has grown enormously in the last three decades. Increasing surface quality, technological developments concerning sophisticated surface diffractometers and synchrotron radiation facilities, and a steady development of surface scattering theory have made this progress possible. Nowadays detailed and precise results from various liquid, glassy, and solid surfaces are available, and even complex layer structures can be characterized.

It is quite clear that there is also a large body of "competing" or complementary methods, such as various kinds of scanning tunneling microscopy, atomic force microscopy (AFM), and high-resolution transmission electron microscopy. Whenever possible, it is desirable to investigate surfaces with such complementary probes to obtain both real-space and reciprocal-space information. There is, however, one major advantage in favour of scattering methods: Soft-matter surfaces, i.e. the surfaces and interfaces of liquids, polymers, glasses, organic multilayers, etc., can be investigated on an atomic scale[1].

In our daily life soft-matter films are playing a more and more important role: Thin polymer films are used as coatings in many technological applications. The variety extends from their simple use as protection of surfaces against corrosion to the use of ultrathin films in semiconductor technology. Organic multilayers are promising materials for biosensors and liquid films are present in so many obvious contexts that there is no need to explain this here.

In addition to the applications and use of soft-matter films in materials science, they are also of interest with respect to fundamental questions: In thin films liquids and polymers may be considered as trapped in a quasi-two-dimensional geometry. This confined geometry is expected to alter the properties and structures of these materials considerably.

On the one hand the typical film thicknesses are on the order of 20–1000 Å, and hence probes for their investigation have to be sensitive to this vertical length scale. Laterally even larger length scales ($> 10\,\mu$m) have to be covered,

[1] It is quite easy to investigate polymer surfaces with AFM. However, the extracted information is most reliable on larger length scales than those accessible by x-ray scattering.

since soft-matter films may possess long-range correlations due to capillary waves on their surfaces. On the other hand many fundamental properties are expected to be dominated by the regions close to the interfaces of the film, i.e. atomic resolution is needed for their investigation. X-ray scattering at grazing angles is a tool which meets these criteria, and hence soft-matter thin films have been extensively investigated by this method in the past. In this book, I review a small fraction of this work, where the focus is on x-ray reflectometry and off-specular diffuse scattering.

The surface sensitivity is based on the fact that for x-ray wavelengths ($\lambda \sim 1\,\text{Å}$) the index of refraction is slightly smaller than unity. Total external reflection occurs if the angle of incidence of the impinging x-rays is sufficiently small – more precisely, smaller than the critical angle of total external reflection. This limits the penetration depth and the scattering to the near surface region. However, this is not quite correct for x-ray reflectometry, since a reflectivity mainly consists of data taken at angles considerably larger than the critical angle, where the penetration depth is already on the order of thousands of angstroms. The sensitivity is then obtained by interference of x-rays scattered at different depths in the sample. In Chap. 2 the theory of x-ray reflectivity is discussed in terms of conventional optics. The basic principles are explained (Fresnel formulas, Parratt algorithm) as well as the more complicated topic of a *strict* inclusion of roughness into the formulas (comparison of "intuitively correct" results with exact results).

The next chapter deals with reflectivity experiments and is divided into two sections: In Sect. 3.1 setups to carry out reflectivity and diffuse-scattering measurements are described. I have taken the liberty of keeping the descriptions concise since nowadays all relevant information may be obtained within minutes by a few mouse clicks via the Internet[2]. Instead, the considerations about including the resolution in the data analysis are presented in more detail: In the specific case of scattering from liquid interfaces many important subtleties have to be taken into account. Section 3.2 contains several examples of soft-matter thin films that have been investigated by x-ray reflectometry. The examples range from the determination of the capillary wave roughness on top of polymer films to density profiles at the liquid/solid interface and structural properties of organic multilayers. Most of these examples again appear in Sect. 6.2 when diffuse-scattering experiments are discussed.

A major defect of x-ray scattering is the well-known phase problem. The data cannot be inverted, since the scattered signal is proportional to the modulus squared of the Fourier transform of the scattering length density, or, in other words, intensities rather than field amplitudes are measured by a detector. The considerations of Chap. 4 reveal that the situation is not as bad as anticipated for the one-dimensional problem of reflectometry. Reflectivity is re-discussed in Sect. 4.1, but within the kinematical theory of scattering.

---

[2] See e.g. *http://www.aps.anl.gov*, *http://www.esrf.fr*, *http://www.nsls.bnl.gov*, *http://www.desy.de/hasylab*, or *http://www.spring8.or.jp*.

It will be shown that for a system consisting of a layer of low density on top of a substrate with a sufficiently large density, the phase of the (complex) reflection coefficient is completely determined by its modulus, i.e. by the measured reflectivity. Fortunately this condition is valid for many soft-matter thin-film systems! A new data inversion method is proposed which is based on this fact and the inclusion of all a-priori knowledge of the system under investigation.

The second part of this book gives a review of a fraction of the works where diffuse x-ray scattering has yielded new information about soft-matter thin films. Before coming to the specific examples, a statistical description of surfaces and interfaces is given in Chap. 5 because the respective height–height correlation functions enter the scattering formulas. The definitions of correlation functions are recapitulated and surfaces with thermally excited capillary waves are considered in great detail. Also some of the formulas which have already appeared in Chap. 2 find their justification here.

In the first part of Chap. 6 (Sect. 6.1) the theory of diffuse x-ray scattering is outlined. I have restricted the description to a simple kinematical treatment (a brief introduction into the so-called distorted-wave Born approximation can be found in Appendix A.2). However, again the subtleties which arise from long-range surface correlations lead to considerable complications in the treatment. In Sect. 6.2 experiments with polymer, liquid, and Langmuir–Blodgett films are presented. Another class of systems consists of polymer films on laterally structured surfaces. These particular substrates enable the quantitative investigation of roughness propagation by soft-matter thin films. Chapter 6 in a sense marks the end of the discussion of "classical" scattering experiments in this book.

A new, promising field for the future is x-ray scattering using coherent radiation. The high-brilliance third-generation synchrotron facilities are able to generate coherent beams with sufficiently high flux for surface investigations. In Chap. 7 a rigorous x-ray scattering theory for the case of (partially) coherent radiation is presented together with some examples of recent measurements. A critical discussion of the possibilities and difficulties is given and the scattering formulas for x-ray photon correlation spectroscopy are explicitly calculated.

At the end of this introduction it may be useful to mention some of the related topics that are not addressed in this book. Here all kinds of *diffraction* from soft-matter thin films have to be mentioned. Thus, diffraction from organic molecules on water surfaces is not addressed. The review of *Als-Nielsen et al.* [12] covers this subject. Grazing-incidence diffraction (GID) in general is not addressed. This method probes surfaces by keeping the incidence and exit angle small while measuring out of the scattering plane, e.g. Bragg reflections. Hence, the scattering is always limited to the first few atomic layers. The reviews of *Dosch* [103] and *Holý, Pietsch & Baumbach* [167] give an extensive overview of GID, where the former concentrates on the determination of critical phenomena in the near-surface region and the latter mainly deals with the structure of multilayers and dynamical scattering theory.

I also tried to avoid recapitulating the whole distorted-wave Born approximation theory of diffuse x-ray scattering, and hence have restricted myself to a description of the simple kinematical treatment. The recent review of *Dietrich & Haase* [94] contains calculations for almost all cases that may occur in experiments.

Finally, it should be mentioned that the field of neutron scattering is not touched in this article [40, 43, 116, 409]. However, all formulas may easily be translated to the case of neutron scattering, if the electron density is replaced by the scattering length density for neutrons. On one hand neutron reflectivity measurements often suffer from very low intensity, but on the other hand there is no better way to study many polymer/polymer and liquid/liquid interfaces yet. The field of neutron reflectivity in connection with soft-matter interfaces has been reviewed by *Russell* [303, 304] and *Stamm* [348], where further details may be found.

2. Reflectivity of X-Rays from Surfaces

The scattering of electromagnetic waves is used in basic research, materials science, and industry for investigations of surfaces and interfaces on length scales covering several orders of magnitude. Radar waves are scattered from the earth and other planets to create detailed maps of the respective surfaces, light scattering is a common technique to investigate the quality of silicon wafers in semiconductor technology on length scales on the order of microns, and x-ray scattering is a probe to investigate surfaces on angstrom scales.

This variety of quite different scattering problems can be described in the same way by introducing a refractive index and solving Maxwell's equations. The particular case of the reflection of hard x-rays (wavelengths on the order of angstroms, photon energies of approximately 10 keV) from matter will be discussed in the following sections.

2.1 Basic Principles

An electromagnetic plane wave given by its electric field vector $\boldsymbol{E}(\boldsymbol{r}) = \boldsymbol{E}_0 \exp(\mathrm{i}\,\boldsymbol{k}_{\mathrm{i}} \cdot \boldsymbol{r})$, which penetrates into a medium characterized by an index of refraction $n(\boldsymbol{r})$, propagates according to the Helmholtz equation

$$\Delta \boldsymbol{E}(\boldsymbol{r}) + k^2 n^2(\boldsymbol{r}) \boldsymbol{E}(\boldsymbol{r}) = 0 , \qquad (2.1)$$

where $k = 2\pi/\lambda$ is the modulus of the wavevector $\boldsymbol{k}_{\mathrm{i}}$ and λ denotes the x-ray wavelength. In general, the index of refraction for an arrangement of N atoms per unit volume, which may be assumed to be harmonic oscillators with resonance frequencies ω_j, is expressed as [178]

$$n^2(\boldsymbol{r}) = 1 + N \frac{e^2}{\varepsilon_0 m} \sum_{j=1}^{N} \frac{f_j}{\omega_j^2 - \omega^2 - 2\mathrm{i}\,\omega\,\eta_j} , \qquad (2.2)$$

where ω is the frequency of the incoming electromagnetic wave, e is the charge and m the mass, respectively, of the electron, the η_j are damping factors, and the f_j denote the forced oscillation strengths of the electrons of each single atom. It should be noted that in general the f_j are complex numbers, $f_j = f_j^0 + f_j'(E) + \mathrm{i} f_j''(E)$, where $f_j'(E)$ and $f_j''(E)$ take into account dispersion and absorption corrections depending on the radiation energy E [394]. For x-rays $\omega > \omega_j$, and Eq. (2.2) may be replaced by

$$n(\boldsymbol{r}) = 1 - \delta(\boldsymbol{r}) + \mathrm{i}\beta(\boldsymbol{r}) , \tag{2.3}$$

with the dispersion and absorption terms

$$\delta(\boldsymbol{r}) = \frac{\lambda^2}{2\pi} r_e \varrho(\boldsymbol{r}) \sum_{j=1}^{N} \frac{f_j^0 + f_j'(E)}{Z} \tag{2.4}$$

and
$$\beta(\boldsymbol{r}) = \frac{\lambda^2}{2\pi} r_e \varrho(\boldsymbol{r}) \sum_{j=1}^{N} \frac{f_j''(E)}{Z} = \frac{\lambda}{4\pi}\mu(\boldsymbol{r}) . \tag{2.5}$$

It should be emphasized that $\delta(\boldsymbol{r})$ is always positive. In Eqs. (2.4) and (2.5) we have introduced the classical electron radius[1] $r_e = e^2/(4\pi\varepsilon_0 mc^2) = 2.814 \times 10^{-5}$ Å; the total number of electrons $Z = \sum_j Z_j$, where Z_j denotes the number of electrons of each component of the material; the electron density $\varrho(\boldsymbol{r})$ as a function of the spatial coordinates $\boldsymbol{r} = (x, y, z) = (\boldsymbol{r}_\parallel, z)$; and the linear absorption coefficient $\mu(\boldsymbol{r})$. The quantities f_j^0 are $\boldsymbol{q}$-dependent, where $\boldsymbol{q} = \boldsymbol{k}_f - \boldsymbol{k}_i$ is the wavevector transfer ($\boldsymbol{k}_i$, $\boldsymbol{k}_f$ are the wavevectors of the incident and scattered plane x-ray waves). This has to be taken into account when measurements over a large q region are analyzed [394]. However, in the region of glancing incident and exit angles, α_i and α_f, respectively, the wavevector transfer is small, and f_j^0 may be approximated with high accuracy by $f_j^0 \approx Z_j$. In the case of a homogeneous medium and far away from absorption edges, one may simplify the expression for the refractive index to

$$n = 1 - \frac{\lambda^2}{2\pi} r_e \varrho + \mathrm{i}\frac{\lambda}{4\pi}\mu . \tag{2.6}$$

Scattering length densities $r_e\varrho$ for some materials are listed in Table 2.1.

The values in Table 2.1 yield $\delta = r_e\varrho\lambda^2/(2\pi) \sim 10^{-6}$ for x-rays, i.e. the real part of the refractive index is slightly smaller than unity. The absorption β is usually one or two orders of magnitude smaller[2].

For a single vacuum/medium interface the law of refraction gives $\cos\alpha_i = (1 - \delta)\cos\alpha_t$, where α_t is the exit angle of the refracted radiation (see Fig. 2.1)[3]. Thus, if $\alpha_t = 0$, and since δ is very small, the critical angle is

$$\alpha_c \approx \sqrt{2\delta} = \lambda\sqrt{r_e\varrho/\pi} . \tag{2.7}$$

---

[1] Also known as the Thompson scattering length of the electron.

[2] For neutrons the refractive index can be written as $n = 1 - \delta_n + \mathrm{i}\beta_n$, too. Whereas δ_n is on the same order as for x-rays, the absorption β_n is essentially negligible ($\beta_n \sim 10^{-12}$).

[3] In the x-ray literature all angles are measured towards the surface and not, as usual in optics, with respect to the surface normal.

Table 2.1. Scattering length densities $r_e\varrho$, dispersions δ, linear absorption coefficients μ, and critical angles α_c for x-rays with $\lambda = 1.54$ Å (Cu$K\alpha$ radiation, from *Stamm* [348] and [394])

	$r_e\varrho\,(10^{10}\mathrm{cm}^{-2})$	$\delta(10^{-6})$	$\mu\,(\mathrm{cm}^{-1})$	$\alpha_c(°)$
Vacuum	0	0	0	0
PS $(C_8H_8)_n$	9.5	3.5	4	0.153
PMMA $(C_5H_8O_2)_n$	10.6	4.0	7	0.162
PVC $(C_2H_3Cl)_n$	12.1	4.6	86	0.174
PBrS $(C_8H_7Br)_n$	13.2	5.0	97	0.181
Quartz (SiO_2)	18.0–19.7	6.8–7.4	85	0.21–0.22
Silicon (Si)	20.0	7.6	141	0.223
Nickel (Ni)	72.6	27.4	407	0.424
Gold (Au)	131.5	49.6	4170	0.570

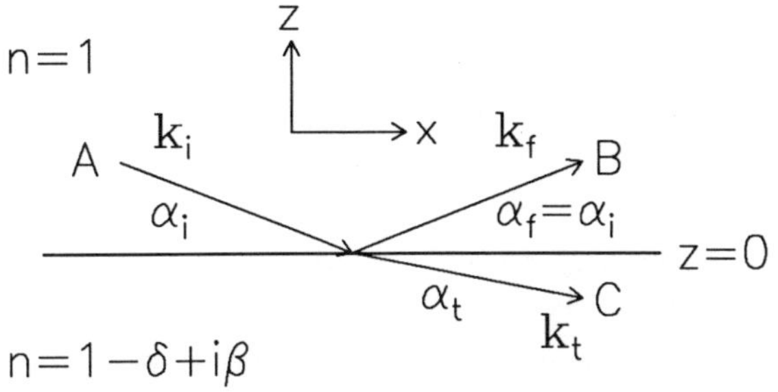

Fig. 2.1. A plane electromagnetic wave with wavevector $\boldsymbol{k}_i$ hits a surface at a grazing angle α_i. The wave splits into a reflected ($\alpha_f = \alpha_i$) and a refracted wave transmitted at the angle α_t

For incident angles $\alpha_i \le \alpha_c$ the phenomenon of total external reflection occurs. The x-rays do not penetrate far into the medium. Instead, all incoming radiation is reflected (with small losses due to absorption). The critical angle α_c is usually several tenths of a degree for most materials (see Table 2.1).

Now the reflectivity of a single perfectly smooth vacuum/medium interface will be calculated and the situation shown in Fig. 2.1 is considered. A plane wave in vacuum, $\boldsymbol{E}_i(\boldsymbol{r}) = (0, A, 0), \exp(\mathrm{i}\,\boldsymbol{k}_i \cdot \boldsymbol{r})$ with wavevector $\boldsymbol{k}_i = k(\cos\alpha_i, 0, -\sin\alpha_i)$, hits at a grazing angle α_i a flat surface of a medium with refractive index $n = 1 - \delta + \mathrm{i}\beta$. The reflected and transmitted fields may be described by $\boldsymbol{E}_r(\boldsymbol{r}) = (0, B, 0)\exp(\mathrm{i}\,\boldsymbol{k}_f \cdot \boldsymbol{r})$, with $\boldsymbol{k}_f = k(\cos\alpha_i, 0, \sin\alpha_i)$, and $\boldsymbol{E}_t(\boldsymbol{r}) = (0, C, 0)\exp(\mathrm{i}\,\boldsymbol{k}_t \cdot \boldsymbol{r})$, where the components of $\boldsymbol{k}_t = (k_{t,x}, 0, k_{t,z})$ are given by the law of refraction. Here the case of s-polarization is considered, i.e. the impinging wave is linearly polarized with the electric field vector in the y direction perpendicular to the (x, z) scattering plane. From the fact that the tangential components of the electric and magnetic fields have to be continuous at the surface $z = 0$, the (complex) reflection and transmission

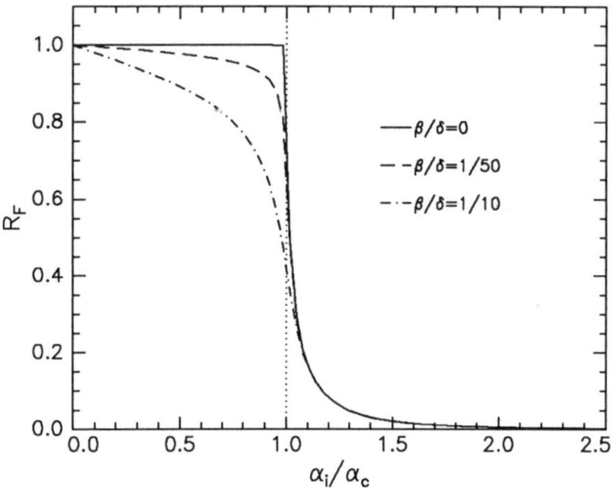

Fig. 2.2. Fresnel reflectivity R_F vs. the angle of incidence α_i normalized by the critical angle α_c of a silicon/vacuum interface ($\alpha_c = 0.22°$) for the wavelength $\lambda = 1.54\,Å$ and different absorption/dispersion ratios β/δ. The absorption is only of importance in the region of the critical angle

coefficients $r_s = B/A$ and $t_s = C/A$ follow. One obtains the well-known Fresnel formulas [44]

$$r_s = \frac{k_{i,z} - k_{t,z}}{k_{i,z} + k_{t,z}} , \tag{2.8}$$

$$t_s = \frac{2\,k_{i,z}}{k_{i,z} + k_{t,z}} , \tag{2.9}$$

where $k_{i,z} = k \sin\alpha_i$ and $k_{t,z} = nk\sin\alpha_t = k(n^2 - \cos^2\alpha_i)^{1/2}$. In the case of p-polarized waves one gets the results [44]

$$r_p = \frac{n^2 k_{i,z} - k_{t,z}}{n^2 k_{i,z} + k_{t,z}} , \tag{2.10}$$

$$t_p = \frac{2\,k_{i,z}}{n^2 k_{i,z} + k_{t,z}} . \tag{2.11}$$

Since n is almost unity for x-rays, in practice there is no difference between the two cases. Therefore, only the s-polarization will be considered throughout this article and the subscript "s" is omitted in the following.

The intensity $R_F = |r|^2$ of the reflected x-ray wave (the so-called Fresnel reflectivity) is given explicitly in the small-angle regime by the expression

$$R_F = \frac{(\alpha_i - p_+)^2 + p_-^2}{(\alpha_i + p_+)^2 + p_-^2} , \tag{2.12}$$

with

$$p_{+/-}^2 = \frac{1}{2}\left\{\sqrt{(\alpha_i^2 - \alpha_c^2)^2 + 4\beta^2} \pm (\alpha_i^2 - \alpha_c^2)\right\} \tag{2.13}$$

being the real and imaginary part of the (complex!) transmission angle $\alpha_t = p_+ + ip_-$.

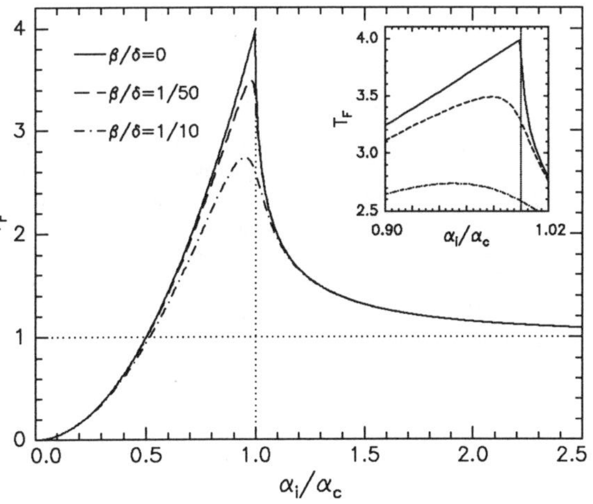

Fig. 2.3. Fresnel transmission T_F as a function of the normalized angle of incidence α_i/α_c for a silicon/vacuum interface, different absorption/dispersion ratios β/δ, and the wavelength $\lambda = 1.54$ Å. The enhanced transmission for $\alpha_i \approx \alpha_c$ can be seen. The inset displays a magnification of the region around the critical angle

Figure 2.2 shows the Fresnel reflectivity (Eq. 2.12) as a function of the incident angle α_i normalized by the critical angle α_c for a fixed value of $\delta = 7.56 \times 10^{-6}$ (silicon, CuKα radiation) and different ratios β/δ. The absorption only plays a role in the vicinity of the critical angle, leading to a "rounding" in this region, and is essentially negligible for larger angles of α_i. The function R_F decreases rapidly for incident angles $\alpha_i > \alpha_c$. An expansion of Eq. (2.12) shows that for $\alpha_i > 3\,\alpha_c$ the Fresnel reflectivity may be well approximated by $R_F \simeq (\alpha_c/2\alpha_i)^4$. This is the reason why x-ray reflectivity measurements always have to cover a dynamic intensity range of more than six orders of magnitude, and in general they are shown on logarithmic scales.

Figure 2.3 depicts the Fresnel transmission $T_F = |t|^2$ (see Eq. 2.9) again as a function of the normalized incident angle for various β/δ ratios. For $\alpha_i \approx \alpha_c$ this function has a pronounced maximum[4], originating from the fact that the reflected and transmitted waves interfere constructively, thus enhancing the amplitude of the transmitted wave by a factor of two[5]. Absorption again damps this enhancement considerably. For larger angles of α_i the Fresnel transmission tends to unity. The impinging wave field penetrates almost unhindered into the medium (see Sect. 4.1). The Fresnel transmission plays an important role in the explanation of the properties of the diffusely scattered intensity. This will be discussed later in Chap. 6.

The last point which will be discussed in this basic introductory section is the penetration depth of x-rays in the small-angle regime. With the complex

[4] The inset of Fig. 2.3 shows that this maximum shifts to slightly smaller values than α_c with increasing absorption.

[5] Here it is important to note that a calculation of the Poynting vector would show that the conservation of energy is – of course – maintained.

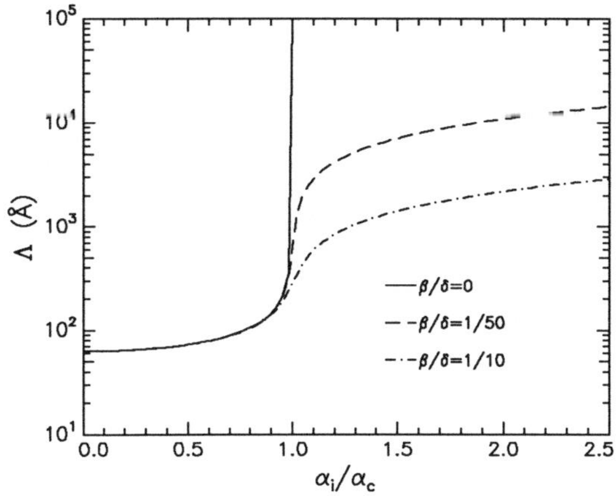

Fig. 2.4. Penetration depth Λ of x-rays ($\lambda = 1.54\,\text{Å}$) into silicon as a function of the normalized angle of incidence $\alpha_\mathrm{i}/\alpha_\mathrm{c}$ for different absorption/dispersion ratios β/δ. Note that without absorption ($\beta = 0$) the penetration depth is not limited for $\alpha_\mathrm{i} > \alpha_\mathrm{c}$. Otherwise, the penetration depth may be approximated for $\alpha_\mathrm{i} \gg \alpha_\mathrm{c}$ by $\Lambda \approx \alpha_\mathrm{i}\lambda/(2\pi\beta)$

angle of refraction, $\alpha_\mathrm{t} = p_+ + \mathrm{i}p_-$, the modulus of the electric field $\boldsymbol{E}_\mathrm{t}$ in the medium ($z \leq 0$) may be expressed by

$$E_\mathrm{t} = |C| \exp\left\{\mathrm{i}(k_{\mathrm{i},x}x - k\,z\,p_+)\right\} \exp(k\,z\,p_-)\,, \tag{2.14}$$

where p_+ and p_- are given by Eq. (2.13). If $\alpha_\mathrm{i} \leq \alpha_\mathrm{c}$, then p_- is rather large and Eq. (2.14) describes an exponentially decaying wave ("evanescent wave"), which travels parallel to the surface. The penetration depth Λ of this wave is given by

$$\Lambda = 1/(kp_-) = \frac{\lambda}{\sqrt{2}\pi}\left\{\sqrt{(\alpha_\mathrm{i}^2 - \alpha_\mathrm{c}^2)^2 + 4\beta^2} - (\alpha_\mathrm{i}^2 - \alpha_\mathrm{c}^2)\right\}^{-1/2}. \tag{2.15}$$

For $\alpha_\mathrm{i} \to 0$ the penetration depth tends to $\Lambda_0 = \lambda/(2\pi\alpha_\mathrm{c}) = 1/\sqrt{4\pi r_\mathrm{e}\varrho}$, thus being independent of the wavelength λ. For most materials one gets $\Lambda_0 \sim 50\,\text{Å}$, which shows that in the region of very small incident angles the scattering mainly stems from the near-surface region. This fact is extensively used in the so-called grazing-incidence-diffraction (GID) technique, where Bragg scattering is combined with the surface sensitivity in the region of the critical angle [101, 102]. Since this is not of importance for reflectivity measurements, the interested reader is referred to the review of *Dosch* [103] and the book of *Holý, Pietsch & Baumbach* [167]. For $\alpha_\mathrm{i} > \alpha_\mathrm{c}$ the penetration depth Λ rapidly increases and is limited only by the absorption of the material. The maximum penetration depth Λ_max follows from Eq. (2.15) as $\Lambda_\mathrm{max} = \lambda/(4\beta) \sim 10^4$–$10^5\,\text{Å}$ for most materials and perpendicular incidence ($\alpha_\mathrm{i} = \pi/2$). Figure 2.4 shows Λ as a function of $\alpha_\mathrm{i}/\alpha_\mathrm{c}$ for different ratios β/δ.

To end this section, an experimentally observed reflectivity is discussed. In 1985 *Braslau et al.* [47, 48] measured the reflected x-ray intensity from a water surface. The result is shown in Fig. 2.5, where the reflectivity R is plotted as a function of $q_z = 2k \sin\alpha_\mathrm{i}$, the perpendicular component of the

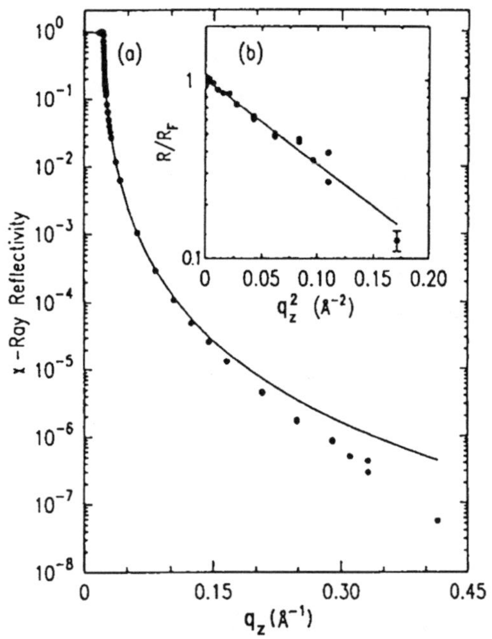

Fig. 2.5. **(a)** Reflectivity R of a water/air interface (*solid circles*) vs. perpendicular wavevector transfer $q_z = 2k \sin \alpha_{\mathrm{i}}$. The *solid line* represents the Fresnel reflectivity R_{F} of a perfectly sharp interface. The pure Fresnel reflectivity is able to explain the measurement over five orders of magnitude. However, R_{F} is always above the data and differences between R_{F} and the measurement are clearly visible for $q_z > 0.15\,\text{Å}^{-1}$. **(b)** The inset shows $\log(R/R_{\mathrm{F}})$ vs. q_z^2, yielding a capillary wave roughness of $\sigma = (3.3 \pm 0.1)\,\text{Å}$, since roughness damps the measured intensity according to $\exp(-\sigma^2 q_z^2)$ (see Sect. 2.3; figure taken from *Braslau et al.* [47, 48])

wavevector transfer. The data (solid circles) are well explained over five orders of magnitude by the Fresnel reflectivity of Eq. (2.12) (solid line), showing a rapid decrease for incident angles $\alpha_{\mathrm{i}} > \alpha_{\mathrm{c}}$. However, for $q_z > 0.15\,\text{Å}^{-1}$ the measured intensity is considerably smaller than the Fresnel reflectivity. The reason is the surface roughness caused by thermally excited capillary waves on the water surface. Surface roughness, particularly capillary waves, will be discussed later (see Sects. 2.3 and 5.3.3). First a simple extension to layered systems will be given in the next section.

2.2 Multiple Interfaces

For practical applications the case of stratified media is much more important than that of a single surface. In almost all cases layer systems are present where the scattering from all interfaces has to be taken into account.

Hence, scattering from several sharp interfaces is considered. Figure 2.6 schematically shows a multilayer stack consisting of N interfaces at the positions $z_j \le 0$. The vacuum counts as "layer 1" with the first interface at $z_1 = 0$. The last interface is located at z_N with the underlying semi-infinite substrate ("layer $N+1$"). The refractive index of each layer, with thickness $d_j = z_{j-1} - z_j$, is denoted by $n_j = 1 - \delta_j + \mathrm{i}\beta_j$, $\boldsymbol{k}_{\mathrm{i},j}$ and T_j are the wavevector and the amplitude of the transmitted wave and $\boldsymbol{k}_{\mathrm{f},j}$ and R_j are

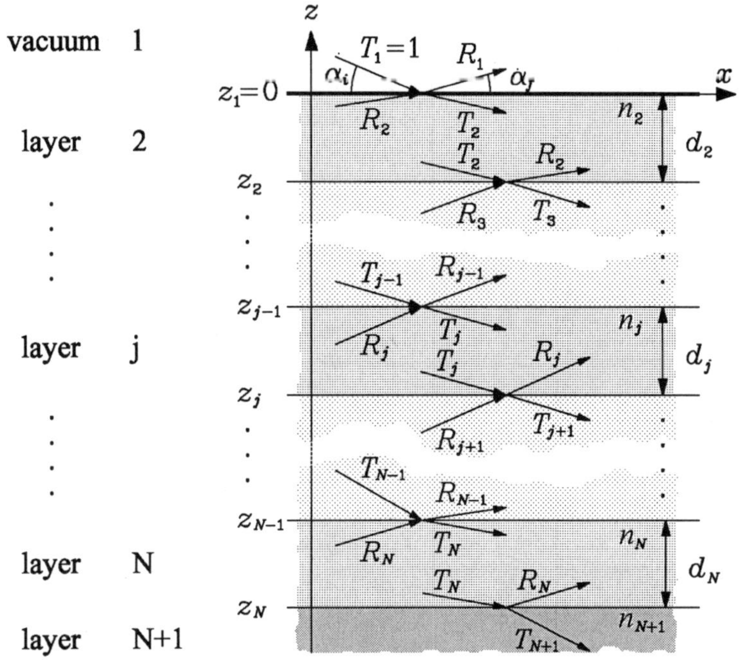

Fig. 2.6. Sketch of a system consisting of $N+1$ layers with N interfaces. In the case of reflectivity the condition $\alpha_i = \alpha_f$ holds. The incident wave amplitude is normalized to unity, $T_1 = 1$. No wave is reflected from the substrate, i.e. $R_{N+1} = 0$

the corresponding values for the reflected wave inside layer j. The impinging wave with an amplitude normalized to unity, $T_1 = 1$, hits the surface at the grazing angle α_i. In what follows, it will be shown how the amplitude R_1 of the specularly reflected wave may be calculated. In so doing one only has to bear in mind that the tangential components of the electric and magnetic field vectors have to be continuous at the interfaces (see Fig. 2.6).

The first theoretical treatment of x-ray reflectivity was given by *Picht* [283] in the year 1929. The dynamical calculation of R_1 was first published in 1950 by *Abelès* [1], who introduced a transfer matrix for each layer, which essentially connects the fields at interface j with those at interface $j+1$, thus yielding the reflected amplitude R_1 after N matrix multiplications (for details see Refs. [1, 35, 44, 206]). Equivalent to this formalism is a recursive approach first described in the year 1954 in the classic paper of *Parratt* [277][6]. If X_{j+1} denotes the ratio of R_{j+1} and T_{j+1} in layer $j+1$, then X_j for the layer above may be calculated via

$$X_j = \frac{R_j}{T_j} = \exp(-2i\, k_{z,j} z_j)\, \frac{r_{j,j+1} + X_{j+1} \exp(2i\, k_{z,j+1} z_j)}{1 + r_{j,j+1} X_{j+1} \exp(2i\, k_{z,j+1} z_j)}\,, \tag{2.16}$$

[6] This is probably the most cited work in the field of x-ray reflectometry!

where

$$r_{j,j+1} = \frac{k_{z,j} - k_{z,j+1}}{k_{z,j} + k_{z,j+1}} \tag{2.17}$$

is the Fresnel coefficient of interface j (see Eq. 2.8) with $k_{z,j} = k(n_j^2 - \cos^2\alpha_{\rm i})^{1/2}$ being the z component of the wavevector in layer j. In general, the substrate is much thicker than the penetration depth of x-rays (see Eq. 2.15). Consequently there is no reflection from the substrate, i.e. one may set $R_{N+1} = X_{N+1} = 0$ as the start of the recursion. The specularly reflected intensity R is obtained from Eq. (2.16) after N iterations:

$$R = \left|X_1\right|^2 = \left|R_1\right|^2 . \tag{2.18}$$

With the knowledge of R_1 and $T_1 = 1$ the amplitudes R_j and T_j inside all layers are given recursively by

$$R_{j+1} = \frac{1}{t_{j+1,j}} \Big\{ T_j\, r_{j+1,j} \exp\Big[- {\rm i}(k_{z,j+1} + k_{z,j}) z_j \Big] + R_j \exp\Big[- {\rm i}(k_{z,j+1} - k_{z,j}) z_j \Big] \Big\} , \tag{2.19}$$

$$T_{j+1} = \frac{1}{t_{j+1,j}} \Big\{ T_j \exp\Big[{\rm i}(k_{z,j+1} - k_{z,j}) z_j \Big] + R_j\, r_{j+1,j} \exp\Big[{\rm i}(k_{z,j+1} + k_{z,j}) z_j \Big] \Big\} , \tag{2.20}$$

with the Fresnel transmission coefficient $t_{j+1,j} = 1 + r_{j+1,j}$ of interface j (see Eq. 2.9). In the next section, Eqs. (2.19) and (2.20) are used for the inclusion of the interface roughness[7].

As an example Fig. 2.7 shows a calculated reflectivity of a polystyrene (PS) film of thickness $d = 800$ Å on a silicon substrate. The wavelength used is $\lambda = 1.54$ Å, where $\delta_{\rm Si} = 7.56 \times 10^{-6}$, $\beta_{\rm Si}/\delta_{\rm Si} = 1/40$, $\delta_{\rm PS} = 3.5 \times 10^{-6}$, and $\beta_{\rm PS}/\delta_{\rm PS} = 1/200$. Since the less dense material is on top, for very small incident angles two critical angles[8], one at $\alpha_{\rm i} = \sqrt{2\delta_{\rm PS}} = 0.15°$ and the other at $\alpha_{\rm i} = \sqrt{2\delta_{\rm Si}} = 0.22°$ (see inset of Fig. 2.7), can be seen[9]. For larger incident angles, $\alpha_{\rm i} > \alpha_{\rm c,Si}$, the reflectivity decreases rapidly according to the $(\alpha_{\rm c}/2\alpha_{\rm i})^4$ law (see Sect. 2.1), but with oscillations superimposed ("Kiessig-fringes" [188]). These oscillations stem from interferences of the reflected waves at the vacuum/PS and PS/silicon interfaces. From the period $\Delta\alpha_{\rm i}$

---

[7] In Appendix A.2 they are also needed for the calculation of the diffuse scattering.

[8] A nice example, where this effect is used for the study of surface phase transitions, was recently given by *Klemradt et al.* [191].

[9] It is worth noting that in the reverse case, i.e. when a dense material is on top of a less dense material, only one critical angle would appear.

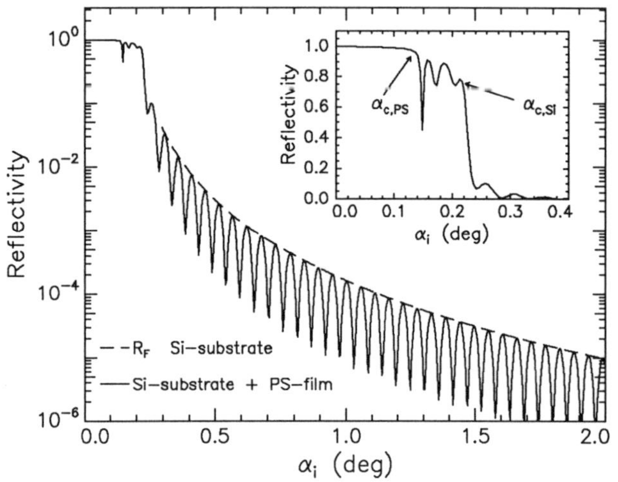

Fig. 2.7. Calculated x-ray reflectivity ($\lambda = 1.54$ Å) of a polystyrene (PS) film of thickness $d = 800$ Å (*solid line*) and Fresnel reflectivity R_F of the substrate (*dashed line* for $\alpha_i > 0.3°$). For clarity, a magnification of the region $\alpha_i < 0.4°$ on a linear scale is shown in the inset, where the two critical angles of PS and Si are visible

the layer thickness may be estimated via $d = 2\pi/\Delta q_z \approx \lambda/(2\Delta\alpha_i)$, where $q_z = 2k \sin \alpha_i$ is the wavevector transfer as defined before.

2.3 Roughness

In practice, surfaces and interfaces are always rough. We have seen in Fig. 2.5 that the measured x-ray reflectivity of a water surface is significantly lower than the Fresnel reflectivity R_F for wavevector transfers $q_z > 0.15$ Å^{-1}. This discrepancy is caused by the roughness of the air/water interface due to thermally induced capillary waves [47, 48].

Up to now in this book, the reflectivity for sharp interfaces has been calculated. Mathematically, this was taken into account by a constant index of refraction n_j which jumps to another constant value n_{j+1} at the boundary between layer j and $j+1$. For a rough surface or interface this sharp step has to be replaced by a continuous variation of the refractive index $n_j(x, y, z)$, i.e. by a continuous electron density $\varrho_j(x, y, z)$. Since here the specular reflectivity is of interest, the wavevector transfer $\boldsymbol{q} = \boldsymbol{k}_f - \boldsymbol{k}_i$ has only a z component. Hence, the structure laterally averaged over (x, y) of a sample is probed. Thus, one-dimensional refractive-index profiles

$$n_j(z) = \iint n_j(x, y, z)\, dy\, dx \tag{2.21}$$

are considered in the following. Lateral inhomogeneities give rise to off-specular diffuse-scattering, which will be discussed in Chap. 6.

The influence of roughness on the specularly reflected intensity may be estimated by averaging Eqs. (2.19) and (2.20) in the direction perpendicular

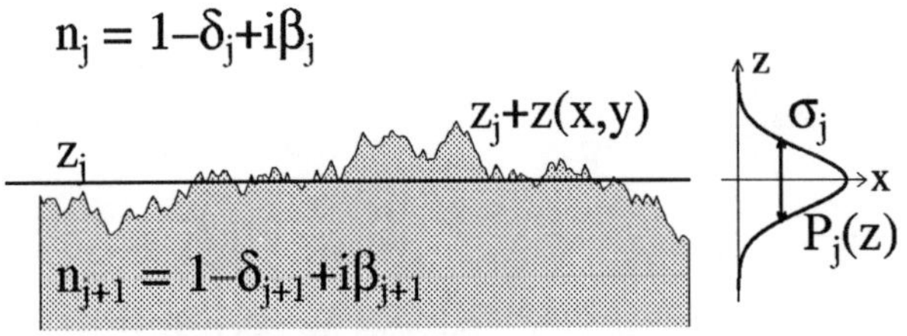

Fig. 2.8. Rough interface with mean z coordinate z_j and fluctuations $z(x, y)$ around this value. The surface is replaced by an ensemble of smooth surfaces at coordinates $z_j + z$ with probability density $P_j(z)$

to the surface [350, 351]. In so doing it is assumed that a rough interface may be replaced by an ensemble of smooth interfaces[10] with certain z coordinates, $z_j + z$, weighted by a probability density $P_j(z)$ (see Fig. 2.8) with mean value

$$\mu_j = \int z\, P_j(z)\, \mathrm{d}z\ , \tag{2.22}$$

and root-mean-square (rms) roughness

$$\sigma_j^2 = \int (z - \mu_j)^2\, P_j(z)\, \mathrm{d}z\ . \tag{2.23}$$

With the function $f_j(k)$ defined by[11]

$$\begin{aligned} f_j(k) &= \Big\langle \exp\Big\{ -\mathrm{i}\, k(z-\mu_j)\Big\}\Big\rangle_{P_j(z)} \\ &= \exp(\mathrm{i}\, k\mu_j) \int \exp(-\mathrm{i}\, kz)\, P_j(z)\, \mathrm{d}z\ , \end{aligned} \tag{2.24}$$

one finds, after averaging the right-hand sides of Eqs. (2.19) and (2.20),

$$\begin{aligned} R_{j+1} = \frac{1}{\tilde{t}_{j+1,j}} \Big\{ & T_j\, \tilde{r}_{j+1,j}\, \exp\Big[-\mathrm{i}(k_{z,j+1} + k_{z,j}) z_j \Big] \\ & + R_j\, \exp\Big[-\mathrm{i}(k_{z,j+1} - k_{z,j}) z_j \Big] \Big\}\ , \end{aligned} \tag{2.25}$$

$$\begin{aligned} T_{j+1} = \frac{1}{f_\mathrm{t}\, \tilde{t}_{j+1,j}} \Big\{ & T_j\, \exp\Big[\mathrm{i}(k_{z,j+1} - k_{z,j}) z_j \Big] \\ & + R_j\, f_\mathrm{r}\, \tilde{r}_{j+1,j}\, \exp\Big[\mathrm{i}(k_{z,j+1} + k_{z,j}) z_j \Big] \Big\}\ , \end{aligned} \tag{2.26}$$

[10] Note that in the following z is used in two slightly different manners: (i) with respect to z_j if $P_j(z)$ is considered, and (ii) with respect to the topmost surface, i.e. $z_1 = 0$ in Fig. 2.6, for the profiles $n_j(z)$.

[11] For real k values $f_j(k)$ is just the Fourier transform of $P_j(z)$.

with the respective modified Fresnel coefficients for rough interfaces

$$\tilde{r}_{j+1,j} = \frac{f_j(k_{z,j+1}+k_{z,j})}{f_j(k_{z,j+1}-k_{z,j})} r_{j+1,j} , \tag{2.27}$$

$$\tilde{t}_{j+1,j} = \frac{1}{f_j(k_{z,j+1}-k_{z,j})} t_{j+1,j} . \tag{2.28}$$

Here $r_{j+1,j}$ and $t_{j+1,j}$ are those for smooth interfaces as calculated before, and

$$f_\mathrm{r} = \frac{f_j(k_{z,j+1}-k_{z,j})}{f_j(-k_{z,j+1}+k_{z,j})} \frac{f_j(-k_{z,j+1}-k_{z,j})}{f_j(k_{z,j+1}+k_{z,j})} , \tag{2.29}$$

$$f_\mathrm{t} = \frac{f_j(k_{z,j+1}-k_{z,j})}{f_j(-k_{z,j+1}+k_{z,j})} . \tag{2.30}$$

Note that $|f_\mathrm{r}|, |f_\mathrm{t}| = 1$ if the arguments are real numbers. In general, the $k_{z,j}$ components may have rather large imaginary parts in the region of the critical angle where absorption is of importance (see Fig. 2.2). However, it turns out that indeed f_r and f_t can be set to unity in very good approximation for the case of x-rays.

If between layers j and $j+1$ a continuous refractive-index profile

$$n_j(z) = \frac{n_j + n_{j+1}}{2} - \frac{n_j - n_{j+1}}{2} \operatorname{erf}\left(\frac{z - z_j}{\sqrt{2}\sigma_j}\right) \tag{2.31}$$

is assumed[12], with the error function defined by

$$\operatorname{erf}(z) = \frac{2}{\sqrt{\pi}} \int_0^z \exp(-t^2)\,\mathrm{d}t , \tag{2.32}$$

a Gaussian probability density ($\mu_j = 0$)

$$P_j(z) = \frac{1}{\sqrt{2\pi}\sigma_j} \exp\left(-\frac{z^2}{2\sigma_j^2}\right) \tag{2.33}$$

results, finally yielding

$$\tilde{r}_{j,j+1} = r_{j,j+1} \exp(-2\,k_{z,j}k_{z,j+1}\sigma_j^2) , \tag{2.34}$$

$$\tilde{t}_{j,j+1} = t_{j,j+1} \exp\left\{ + (k_{z,j} - k_{z,j+1})^2\sigma_j^2/2\right\} , \tag{2.35}$$

for the modified Fresnel coefficients. It should be noted that in the case of a single surface these coefficients directly describe the reflectivity and transmittivity of that interface.

Simple mean-field theoretical arguments would suggest that the refractive index of a liquid/vapor interface follows a hyperbolic tangent[13]. The volume

[12] In the case $\sigma_j \to 0$ one has $\operatorname{erf}[(z - z_j)/(\sqrt{2}\sigma_j)] \to \pm 1$ for $z > z_j$ and $z < z_j$, respectively, thus regaining the sharp-interface case.

[13] An improvement over this obtained from more sophisticated theories is the *Fisk–Widom* [124] profile, where in Eq. (2.36) $\tanh(z)$ is replaced by $\sqrt{2}\tanh(z/2)/[3-\tanh^2(z/2)]^{1/2}$.

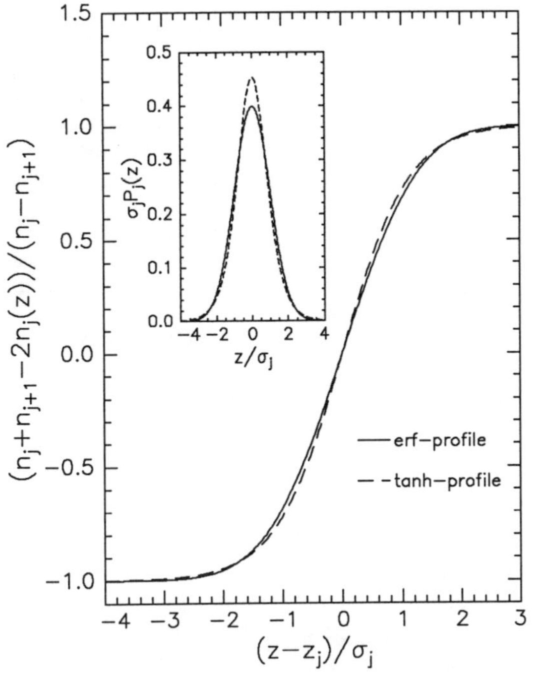

Fig. 2.9. Normalized error-function (*solid line*) and hyperbolic-tangent (*dashed line*) refractive-index profiles $n_j(z)$. The inset shows the respective probability densities $\sigma_j\, P_j(z)$ normalized to the rms roughness. The two profiles are very similar with the difference always being less than 5%. The probability density corresponding to the error-function profile is $P_j(z) \propto \exp(-z^2)$, and that corresponding to the tanh profile is $P_j(z) \propto \cosh^{-2}(z)$. The two probability densities are symmetric with zero mean value μ_j

fraction profile of one component for an interface between two immiscible polymers is predicted to take a hyperbolic-tangent function form, too [36, 50, 157, 158, 325, 326, 349, 363]. Hence, another important symmetric profile for soft-matter interfaces is given by

$$n_j(z) = \frac{n_j + n_{j+1}}{2} - \frac{n_j - n_{j+1}}{2} \tanh\left(\frac{\pi}{2\sqrt{3}} \frac{z - z_j}{\sigma_j}\right) , \tag{2.36}$$

with the probability density ($\mu_j = 0$)

$$P_j(z) = \frac{\pi}{4\sqrt{3}\,\sigma_j} \cosh^{-2}\left(\frac{\pi}{2\sqrt{3}} \frac{z}{\sigma_j}\right) , \tag{2.37}$$

which leads to

$$\tilde{r}_{j,j+1} = \frac{\sinh\left[\sqrt{3}\,\sigma_j\,(k_{z,j} - k_{z,j+1})\right]}{\sinh\left[\sqrt{3}\,\sigma_j\,(k_{z,j} + k_{z,j+1})\right]} , \tag{2.38}$$

$$\tilde{t}_{j,j+1} = t_{j,j+1}\, \frac{\sinh\left[\sqrt{3}\,\sigma_j\,(k_{z,j} - k_{z,j+1})\right]}{\sqrt{3}\,\sigma_j\,(k_{z,j} - k_{z,j+1})} . \tag{2.39}$$

Figure 2.9 shows the two profiles and probability densities. The difference is rather small and always less than 5%. These profiles are often used in the literature to explain reflectivity data. It is important to notice that $|\tilde{r}_{j,j+1}|$ is always smaller than $|r_{j,j+1}|$, i.e. roughness damps the specularly reflected

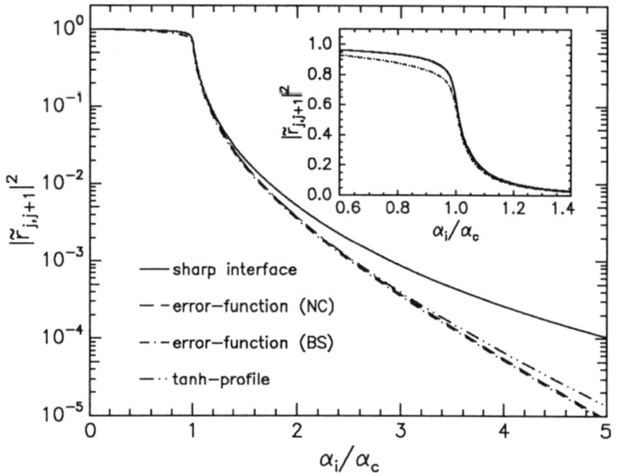

Fig. 2.10. X-ray ($\lambda = 1.54$ Å) reflectivity vs. the normalized incidence angle α_i/α_c of a silicon surface with rms roughness $\sigma_j = 10$ Å. The inset displays a magnification of the region of the critical angle α_c on a linear scale. (NC=*Névot–Croce*, BS=*Beckmann–Spizzichino*, see text)

intensity as can be seen in Fig. 2.10. The reflection coefficients given by Eqs. (2.34) and (2.38) differ slightly for large incident angles and are essentially the same in the region of the critical angle [17, 18]. For the transmission the reverse situation is found: The two profiles yield $|\tilde{t}_{j,j+1}| > |t_{j,j+1}|$ (see Fig. 2.11). The $|\tilde{t}_{j,j+1}|$ curves are almost identical. However, from this finding it cannot be concluded that roughness enhances the transmission of an interface in general. The reason is that the averaging process of Eqs. (2.19) and (2.20) is not strictly justified, since *first* all phase factors were collected at the right-hand side, and then the averages of the exponentials were taken, leading to the $f_j(k_{z,j+1} \pm k_{z,j})$-functions. If these averages were taken *before* all phase factors were collected at the right-hand side[14], then the following modified reflection and transmission coefficients would be obtained for an error-function profile:

$$\tilde{r}_{j,j+1} = r_{j,j+1} \exp(-2\, k_{z,j}^2 \sigma_j^2) \,, \tag{2.40}$$

$$\tilde{t}_{j,j+1} = t_{j,j+1} \exp\left\{ -(k_{z,j} - k_{z,j+1})^2 \sigma_j^2/2 \right\} . \tag{2.41}$$

Whereas the reflection coefficient is almost unaffected and $|\tilde{r}_{j,j+1}| < |r_{j,j+1}|$ still holds, the transmission coefficient has changed drastically and now $|\tilde{t}_{j,j+1}| < |t_{j,j+1}|$. Figures 2.10 and 2.11 depict the reflectivity $|\tilde{r}_{j,j+1}|^2$ and the transmission $|\tilde{t}_{j,j+1}|^2$ of a single rough surface. The results given by Eqs. (2.34)–(2.35) and (2.38)–(2.41) are plotted. The figures reveal that differences between the curves are most prominent in the region of the critical angle, while for larger incidence angles differences are hardly visible.

However, it should be stated here that averaging of Eqs. (2.19) and (2.20) is *not* equivalent to an exact solution of the scattering problem. Thus, basic

[14] This means that four averages $\langle \exp(\pm i k_{z,j} z) \rangle_{P_j(z)}$ for the pre-factors of T_j, R_j and T_{j+1}, R_{j+1} have to be calculated.

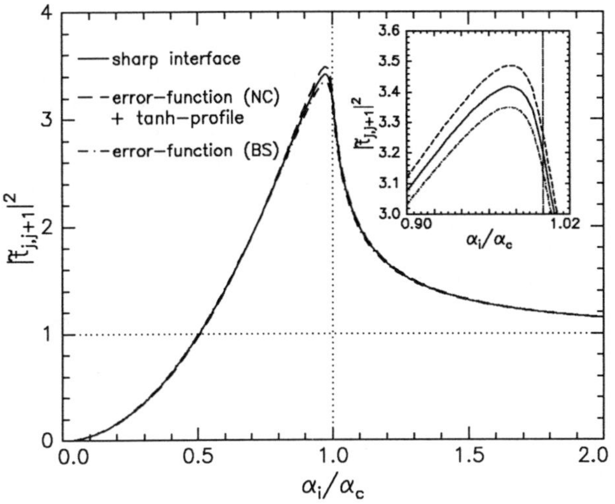

Fig. 2.11. X-ray ($\lambda = 1.54$ Å) transmission $|\tilde{t}_{j,j+1}|^2$ vs. the normalized incidence angle α_i/α_c of a silicon surface with rms roughness $\sigma_j = 10$ Å. The inset shows a magnification of the region of the critical angle α_c (NC=*Névot–Croce*, BS=*Beckmann–Spizzichino*, see text)

conclusions cannot be drawn from this procedure alone. The results have to be carefully compared with calculations from scattering theories based on solutions of the Helmholtz equation (Eq. 2.1) for particular profiles $n(x, y, z)$.

In 1963 *Beckmann & Spizzichino* [31] showed that the reflectivity of a vacuum ($n_1 = 1$)/medium ($n_2 = n$) interface with rms roughness σ is damped by a factor $\exp(-2k_{z,1}^2\sigma^2)$ if "shadowing effects" can be neglected. This means that the local surface curvature is assumed to be much smaller than the wavelength of the impinging radiation. The *Beckmann-Spizzichino* result agrees with that of Eq. (2.40). Later, in 1980, *Névot & Croce* [261] solved Eq. (2.1) for the error-function profile. They published a series of papers (e.g. Refs. [62, 261]), and finally got the result that roughness damps the specularly reflected intensity by a factor $\exp(-2k_{z,1}k_{z,2}\sigma^2)$. This factor agrees with the factor found in Eq. (2.34). However, *Névot & Croce* made approximations which essentially require a locally rapid curvature change of the surface contour in their calculations ("predominance of high spatial frequencies"). Nowadays, the factor $\exp(-2k_{z,j}k_{z,j+1}\sigma_j^2)$ is referred to as the *Névot–Croce* factor in the literature. In 1984, *Vidal & Vincent* [385] confirmed this result. *Sinha et al.* [335] in 1988 and *Pynn* [290] in 1992 calculated the *Névot–Croce* factor in a completely different way. They used the distorted-wave Born approximation (DWBA), a scattering theory that will be outlined later in Sect. 6.1.2 and Appendix A.2. All these calculations indicated that the lateral structure of the roughness decides whether the *Névot–Croce* factor or the *Beckmann–Spizzichino* result is obtained. Although very similar for large $k_{z,1}$, the two results differ remarkably in the region of the critical angle (see inset of Figs. 2.10 and 2.11).

A defect of the work of *Sinha et al.* [335] and *Pynn* [290] is that the scattering cross section was calculated in the first-order DWBA, where the total

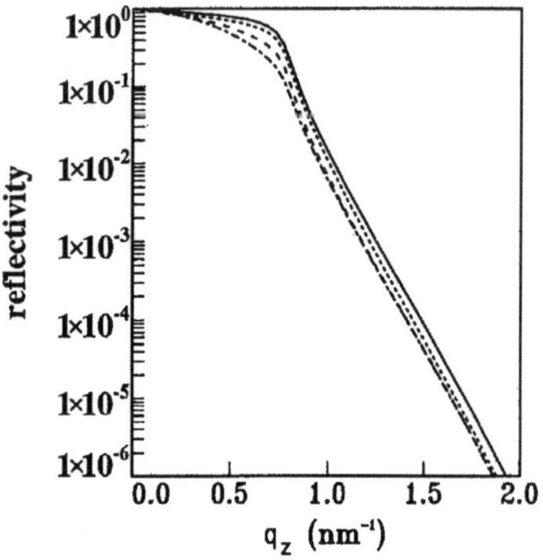

Fig. 2.12. Calculated specular reflectivity vs. wavevector transfer q_z for $\lambda = 1.54\,\text{Å}$. A gold surface with rms roughness $\sigma = 15\,\text{Å}$ is assumed. From *top* to *bottom* the lateral correlation length is $\xi = 0, 1000, 10\,000$, and ∞ Å, corresponding to the *Névot–Croce* ($\xi = 0$) and *Beckmann & Spizzichino* ($\xi = \infty$) results. The region close to the critical angle is strongly affected by the different lateral correlation lengths. For larger q_z values differences are hardly visible (figure taken from *de Boer* [71])

intensity[15] is not conserved. Finally, *de Boer* [71, 72, 73] was able to connect all previously published results: He calculated in 1994 the reflection and transmission coefficients for a single rough surface by applying the DWBA up to second order[16]. The general result for the modified Fresnel coefficients of an error-function profile with width σ (see Eq. 2.31) is

$$\tilde{r} = r\ \exp\Big\{ -2\,k_{z,1}k_{z,2}\big[\sigma^2 - I(\boldsymbol{k}_\parallel)\big]\Big\}\ \exp\Big\{ -2\,k_{z,1}^2 I(\boldsymbol{k}_\parallel)\Big\}\,, \tag{2.42}$$

$$\tilde{t} = t\ \exp\Big\{ +(k_{z,1}-k_{z,2})^2\big[\sigma^2/2 - I(\boldsymbol{k}_\parallel)\big]\Big\}\,. \tag{2.43}$$

Here $\boldsymbol{k} = (\boldsymbol{k}_\parallel, k_{z,1})$ is the wavevector in the vacuum and $k_{z,2}$ is the z component inside the medium as defined before;

$$I(\boldsymbol{k}_\parallel) = \frac{1}{4\pi^2}\,(k_{z,1}+k_{z,2}) \int_{|\boldsymbol{p}_\parallel|<k} \tilde{C}(\boldsymbol{p}_\parallel - \boldsymbol{k}_\parallel)\frac{\mathrm{d}\boldsymbol{p}_\parallel}{p_{z,1}+p_{z,2}}\,, \tag{2.44}$$

where $\boldsymbol{p} = (\boldsymbol{p}_\parallel, p_{z,1})$ is a vector with length $k = 2\pi/\lambda$, and $p_{z,2} = (p_{z,1}^2 - k^2\sin^2\alpha_c)^{1/2}$. The function $\tilde{C}(\boldsymbol{p}_\parallel)$ is the Fourier transform of $C(\boldsymbol{r}_\parallel)$, the height–height correlation function of the rough surface contour. This function $C(\boldsymbol{r}_\parallel)$ can be parametrized quite often by introducing a lateral correlation length ξ. Small ξ values indicate surfaces with rapid local curvature changes whereas surfaces with slowly varying curvature possess very large correlation lengths (see Chap. 5 and Fig. 5.3).

Figure 2.12 depicts the reflectivity given by the modulus squared of Eq. (2.42). If Eqs. (2.42)–(2.44) are evaluated in the limit $k_{z,1}^2\xi/k \ll 1$, one may easily show that $I(\boldsymbol{k}_\parallel) \to 0$, and Eqs. (2.34) and (2.35) are regained, i.e. the *Névot–Croce* result. In the opposite limit, $k_{z,1}^2\xi/k \gg 1$, Eq. (2.44)

[15] This means specular plus diffuse intensity.

[16] In fact, this is the lowest order where the total intensity is conserved.

yields $I(\boldsymbol{k}_{\|}) \to \sigma^2$, thus leading to the *Beckmann–Spizzichino* factors given by Eqs. (2.40) and (2.41). Therefore, the result of *de Boer* is able to explain the reflectivity coefficients which were previously derived, and gives the precise conditions under which the different formulas are applicable. It should also be noted that all results which were obtained by averaging Eqs. (2.19) and (2.20) have been regained, particularly those for the transmission coefficients. Although Eq. (2.42) has not yet been experimentally tested, the fundamental character of the results given by Eqs. (2.42)–(2.44) should be emphasized here. *De Boer* proved that in principle the lateral structure, not only the roughness σ, of an interface has an influence on the specularly reflected intensity. However, in most cases this influence is small and may even be neglected for incident angles larger than the critical angle.

The above-mentioned *de Boer* result was obtained by solving the total scattering problem including the diffuse scattering. However, the reflectivity alone can be treated in a more simple way. The refractive index $n(x, y, z)$ may be expressed by

$$n(x,y,z) = n(z) + \delta n(x,y,z)\,, \tag{2.45}$$

where $n(z)$ is the profile which yields the specular reflectivity and $\delta n(x, y, z)$ accounts for the diffuse scattering[17]. If $\delta n(x, y, z)$ is small, i.e. if the diffuse scattering is negligible compared to the specular reflectivity, then the one-dimensional Helmholtz equation

$$\frac{\mathrm{d}^2 E(z)}{\mathrm{d}z^2} + k_{z,1}^2 n^2(z) E(z) = 0 \tag{2.46}$$

for the electric field $E(z)$ with the respective profile $n(z)$ fully determines the reflectivity[18].

Equation (2.46) can be solved analytically for the tanh profile given by Eq. (2.36). This has been shown by several authors in the past (see e.g. Refs. [15, 206, 207, 397]). A general solution for the electric field is presented in the book by *Lekner* [206] and will be briefly discussed here. This book also gives a discussion of many other profiles for which the Helmholtz equation can be solved exactly.

Without approximations the solution for a vacuum/medium tanh interface of width σ is (note that $j = 1$, $n_1 = 1$, $n_2 = n$)

$$\begin{aligned} E(z) = (-1)^{-\mathrm{i}\,k_{z,2}\sqrt{3}\,\sigma/\pi} &\exp(-\mathrm{i}\,k_{z,2}\,z) \\ \times F\Big[&\mathrm{i}\sqrt{3}\,\sigma\,(k_{z,1} - k_{z,2})/\pi, -\mathrm{i}\sqrt{3}\,\sigma\,(k_{z,1} + k_{z,2})/\pi, \\ &1 - 2\mathrm{i}\sqrt{3}\,\sigma\,k_{z,2}/\pi; -\exp\{\pi z/(\sqrt{3}\,\sigma)\}\Big]\,, \end{aligned} \tag{2.47}$$

with the confluent hypergeometric function [4, 147]

---

17 For an extensive discussion of this ansatz see *Dietrich & Haase* [94] and Chap. 6.

18 Under these conditions the lateral structure does not affect the reflectivity.

$$F(a,b,c;x) = 1 + \frac{a\,b}{c}\frac{x}{1!} + \frac{a(a+1)\,b(b+1)}{c(c+1)}\frac{x^2}{2!} + \dots \,. \tag{2.48}$$

In the far field ($z \to \infty$) Eq. (2.47) reduces to

$$E(z) \longrightarrow A\,\exp(-\mathrm{i}k_{z,1}z) + B\,\exp(\mathrm{i}k_{z,1}z)\;, \tag{2.49}$$

with the reflection coefficient

$$\tilde{r} = \frac{B}{A} = \frac{\sinh\left[\sqrt{3}\,\sigma\,(k_{z,1} - k_{z,2})\right]}{\sinh\left[\sqrt{3}\,\sigma\,(k_{z,1} + k_{z,2})\right]}\; G(\sigma, k_{z,1}, k_{z,2})\;, \tag{2.50}$$

where

$$G(\sigma, k_{z,1}, k_{z,2}) = -\frac{\Gamma(2\sqrt{3}\,\mathrm{i}\,\sigma k_{z,1}/\pi)\;\Gamma[-\sqrt{3}\,\mathrm{i}\,\sigma(k_{z,1}+k_{z,2})/\pi)]}{\Gamma(-2\sqrt{3}\,\mathrm{i}\,\sigma k_{z,1}/\pi)\;\Gamma[\sqrt{3}\,\mathrm{i}\,\sigma(k_{z,1}+k_{z,2})/\pi)]} \times \frac{\Gamma[-\sqrt{3}\,\mathrm{i}\,\sigma(k_{z,1}-k_{z,2})/\pi)]}{\Gamma[\sqrt{3}\,\mathrm{i}\,\sigma(k_{z,1}-k_{z,2})/\pi)]}\;, \tag{2.51}$$

with the Gamma function $\Gamma(x)$ [4, 147]. A comparison of the exact result given by Eq. (2.50) with that of Eq. (2.38) obtained by averaging Eqs. (2.19) and (2.20) shows that the two results are essentially equal, except for the factor $G(\sigma, k_{z,1}, k_{z,2})$. A numerical analysis reveals that $G(\sigma, k_{z,1}, k_{z,2})$ may be replaced by unity in the case of hard x-rays and for roughnesses $\sigma < 100$ Å [153][19]. Hence, in practice Eqs. (2.50) and (2.38) may be considered as identical.

In conclusion, the discussion of the exact results for both the error-function and the tanh profiles reveals that averaging Eqs. (2.19) and (2.20) yields excellent approximations for the reflection and transmission coefficients $\tilde{r}_{j,j+1}$ and $\tilde{t}_{j,j+1}$. Thus, even more complex profiles can be treated analytically in a simple and straightforward way.

The two profiles which were discussed are symmetric with zero mean value $\mu_j = 0$. Many soft-matter interfaces, particularly vapor/liquid boundaries, have been found to be asymmetric [169, 225, 359] (see also Sect. 4.1.1). Therefore, a generalization of the tanh profile is considered,

$$n_j(z) = \frac{n_j + n_{j+1}}{2} - \frac{n_j - n_{j+1}}{2}\, Y\,(a, b; z - z_j)\;, \tag{2.52}$$

which depends on two parameters a and b, with the probability density

$$P_j(z) = \frac{1}{\pi}\frac{(b+a)^2}{b-a}\sin\left(\frac{2\pi\,a}{b+a}\right)\frac{1}{\left[\exp(az) + \exp(-bz)\right]^2}\,. \tag{2.53}$$

The function $Y(a,b;z)$ is explicitly given by

$$Y(a,b;z) = \frac{2}{\pi}\frac{b+a}{b-a}\sin\left(\frac{2\pi\,a}{b+a}\right)\, B_{t(z)}\left(\frac{2b}{b+a}, \frac{2a}{b+a}\right) - 1\;, \tag{2.54}$$

[19] Without absorption ($\beta = 0$) even $|G(\sigma, k_{z,1}, k_{z,2})| \equiv 1$ is valid [153, 206].

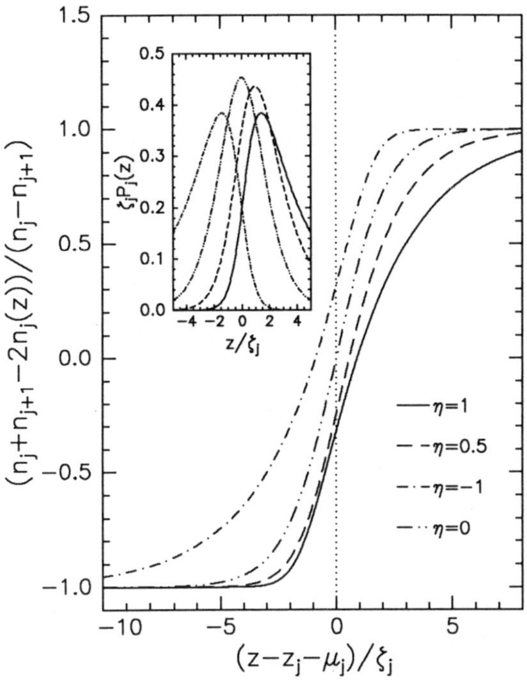

Fig. 2.13. Normalized asymmetric hyperbolic tangent profiles corresponding to the probability density $P_j(z) \propto [\exp(az) + \exp(-bz)]^{-2}$. For notational convenience the parameters a and b have been replaced by means of ζ_j and η, with $a = \pi/(2\sqrt{3}\,\zeta_j)\exp(-\eta)$ and $b = \pi/(2\sqrt{3}\,\zeta_j)\exp(+\eta)$. The case $\eta = 0$ corresponds to a symmetric tanh profile with $\sigma_j = \zeta_j$. Profiles are shown for $\zeta_j = 2$ Å and different asymmetries $\eta = -1, 0, 0.5, 1$. The inset shows the respective normalized probability densities $\zeta_j\, P_j(z)$

with $B_x(\alpha, \beta)$ being the incomplete Beta function as defined in Refs. [4, 147], and

$$t(z) = \frac{1}{1 + \exp[-(b+a)z]} \,. \tag{2.55}$$

Figure 2.13 shows the profile and probability density given by Eqs. (2.52) and (2.53) compared with the symmetric tanh profile. Note that in the case $a = b$ Eqs. (2.52) and (2.53) reduce to Eqs. (2.36) and (2.37) with $\sigma_j = \pi/(2\sqrt{3}\,a)$. The mean value and the rms roughness may be calculated via Eqs. (2.22) and (2.23). One obtains[20]

$$\mu_j = \frac{\pi}{b+a}\cot\left(\frac{2\pi\, a}{b+a}\right) + \frac{1}{b-a} \,, \tag{2.56}$$

$$\sigma_j^2 = \frac{\pi^2}{(b+a)^2}\sin^{-2}\left(\frac{2\pi\, a}{b+a}\right) - \frac{1}{(b-a)^2} \,, \tag{2.57}$$

where the subscript j has been omitted from the parameters a and b. Averaging Eqs. (2.19) and (2.20) with the probability density given by Eq. (2.53) yields

[20] The respective integrals are not listed in Ref. [147] and were calculated by series expansions.

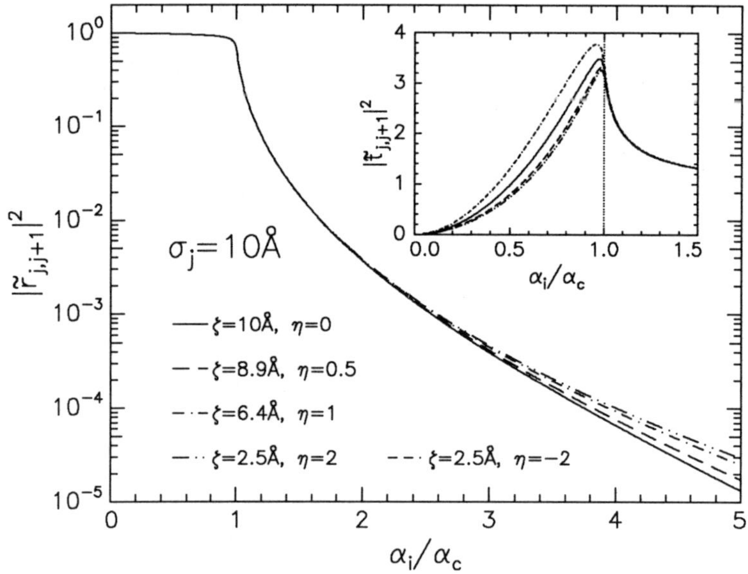

Fig. 2.14. Reflectivity $|\tilde{r}_{j,j+1}|^2$ and transmission $|\tilde{t}_{j,j+1}|^2$ (inset) vs. the normalized incidence angle α_i/α_c for an asymmetric tanh profile (see Fig. 2.13) with rms roughness $\sigma_j = 10$ Å, for different values of the parameters ζ and η

$$\tilde{r}_{j,j+1} = r_{j,j+1}\,\frac{1-\chi^+}{1-\chi^-}\,\frac{\sin(\pi\chi^-)}{\sin(\pi\chi^+)}\,, \tag{2.58}$$

$$\tilde{t}_{j,j+1} = t_{j,j+1}\,\frac{1-(\chi^+ + \chi^-)/2}{1-\chi^-}\,\frac{\sin(\pi\chi^-)}{\sin[\pi(\chi^+ + \chi^-)/2]}\,, \tag{2.59}$$

with

$$\chi^\pm = \frac{2a}{b+a} - \mathrm{i}\,\frac{k_{z,j} \pm k_{z,j+1}}{b+a}\,. \tag{2.60}$$

Figure 2.14 shows the reflection and transmission (inset) coefficients $|\tilde{r}_{j,j+1}|^2$ and $|\tilde{t}_{j,j+1}|^2$, respectively, for different parameters. For notational convenience, instead of a and b the new parameters ζ_j and η have been introduced by means of

$$a = \pi/(2\sqrt{3}\,\zeta_j)\exp(-\eta) \quad \text{and} \quad b = \pi/(2\sqrt{3}\,\zeta_j)\exp(+\eta)\,. \tag{2.61}$$

The asymmetry of the profile is now determined by η ($\eta = 0$: symmetric tanh profile), and ζ_j would be the rms roughness σ_j of the corresponding tanh profile with $a = b$. Figure 2.14 reveals that the reflectivity is slightly affected by the asymmetry[21] of the profile for large incident angles α_i. Whereas it

[21] However, the closer look in Sect. 4.1.1 reveals that the small differences are caused only by the *symmetric* departures of the profile given by Eq. (2.52) from an error-function profile and hence not by the asymmetry.

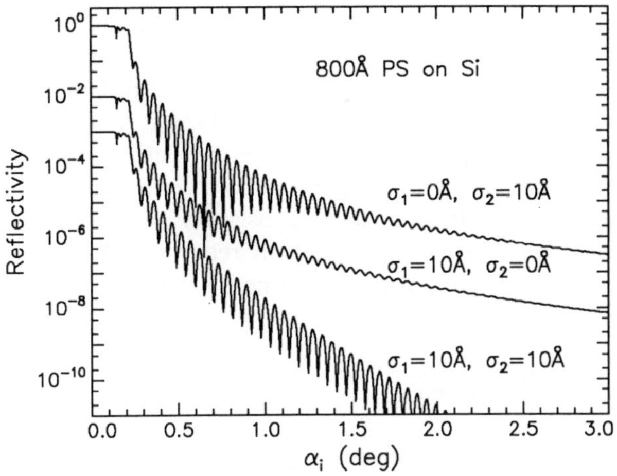

Fig. 2.15. Calculated x-ray reflectivities of a polystyrene (PS) film of thickness $d = 800\,\text{Å}$ on silicon for a wavelength $\lambda = 1.54\,\text{Å}$ and for various rms roughnesses σ_1 (vacuum/PS interface) and σ_2 (PS/silicon interface). The curves are shifted for clarity on the intensity scale

can be proven analytically that $|\tilde{r}_{j,j+1}|^2$ is identical for $+\eta$ and $-\eta$, this does not hold for $|\tilde{t}_{j,j+1}|^2$. The inset of Fig. 2.14 shows that the transmission of the interface differs significantly for the two curves for $\eta = 2$ and $\eta = -2$ in the region $\alpha_\mathrm{i} < \alpha_\mathrm{c}$. However, the discussion of asymmetric refractive index profiles is stopped here since the effects of asymmetries on the reflectivity of a single interface are rather small, and in practice difficult to observe. The reason for this will be clarified in Sect. 4.1, where the reflectivity will be re-discussed in terms of a kinematical theory.

Finally, interface roughness has to be included in the calculation of the reflectivity of a layer system. This can be done quite easily for rms roughnesses which are much smaller than the respective layer thicknesses, i.e. for $\sigma_j \ll d_j$. One simply has to replace in the recursive *Parratt* formalism (Eqs. 2.16 and 2.18) the Fresnel coefficients $r_{j,j+1}$ for sharp interfaces (Eq. 2.17) by the modified coefficients $\tilde{r}_{j,j+1}$ calculated above (Eqs. 2.34, 2.38, 2.40, 2.42, or 2.58).

Again a PS film of thickness $d = 800\,\text{Å}$ is considered as example. The same parameters as used for the calculation of the curves shown in Fig. 2.7 were assumed, and additionally roughnesses σ_1 of the vacuum/PS interface and σ_2 of the PS/Si interface were taken into account via Eq. (2.38) (tanh profile roughness). Figure 2.15 shows the reflectivity for the three cases $\sigma_1 = 0\,\text{Å}$ and $\sigma_2 = 10\,\text{Å}$, $\sigma_1 = 10\,\text{Å}$ and $\sigma_2 = 0\,\text{Å}$, and $\sigma_1 = 10\,\text{Å}$ and $\sigma_2 = 10\,\text{Å}$. Always the condition $\sigma_1, \sigma_2 \ll d$ is fulfilled. It can be seen that roughness essentially damps the modulations stemming from the interference of the x-rays reflected from the substrate and film surfaces. The modulation amplitude remains unaffected only if the roughnesses are identical. But the intensity drops more quickly than in the case of sharp interfaces (see Fig. 2.7). If at least one interface is sharp, i.e. $\sigma = 0$, the general decrease of the intensity is equal to the Fresnel reflectivity, with superimposed damped oscillations. This result can be generalized: For large wavevector transfers $q_z = 2k \sin \alpha_{\rm i}$ the interface with the smallest roughness in a multilayer stack determines the reflectivity. All other contributions from interfaces with larger roughnesses "die" more quickly. According to Eqs. (2.34) or (2.40) these contributions decay exponentially. However, at the end of this section we must repeat the fact that one has to check the condition $\sigma_j \ll d_j$ for *all* layers carefully before applying the formalism given above. If a multilayer consists of interfaces with large roughnesses, this condition does not hold, and the reflectivity has to be calculated in a different way. This is the topic of the next section.

2.4 Arbitrary Density Profiles

Many experiments can be explained with the *Parratt* formalism as introduced in Sect. 2.2 together with the modified Fresnel coefficients $\tilde{r}_{j,j+1}$ of rough interfaces.

We will discuss more quantitatively under which conditions this treatment yields correct results. The upper left panel in Fig. 2.16 depicts a layer stack with sharp interfaces, and the corresponding dispersion profile is on the right (steps δ_j at $z = z_j$). If small roughnesses are present (Fig. 2.16, lower panels), this situation changes slightly. The real dispersion profile $\delta(z)$ shown in the lower right panel of Fig. 2.16 may be approximated by a combination of single error-function or tanh profiles for each interface, still yielding a quasi-continuous model profile. Thus, all layers can be considered as independent and possessing certain interface profiles (error-function, tanh, etc...), leading to the results of the previous sections.

The condition "small roughness" means more precisely that $\sigma_j \ll d_j$ has to be fulfilled to guarantee that the two values n_j and n_{j+1} of the refractive index for layer j and $j+1$ are (almost) reached within these layers. Otherwise the investigated system cannot be treated as a layer stack consisting of N *independent* layers with refractive indices n_j and roughnesses σ_j any more. Figure 2.17a depicts an example: The dispersion profile[22] $\delta(z)$ obtained by the assumption of independent interfaces is not continuous (see Fig. 2.17b). Therefore, the real profile, which of course has to be continuous, and that for which the reflectivity is actually calculated with the *Parratt* formalism, if the simple procedure with the coefficients $\tilde{r}_{j,j+1}$ is applied, differ significantly.

Nevertheless, it is possible to use the *Parratt* formalism for the calculation of the reflectivity when the condition $\sigma_j \ll d_j$ is not fulfilled. In this case a modified procedure has to be adopted: One first has to "guess" the whole profile $\delta(z)$, depending on some free adjustable parameters which later have

[22] Since $n(z) = 1 - \delta(z) + {\rm i}\beta(z)$ one has to consider an absorption profile $\beta(z)$, too. Because both δ and β are proportional to the electron density $\varrho(z)$, it is sufficient to restrict the discussion to $\delta(z)$ in the following. Moreover, for soft-matter thin films absorption is almost negligible.

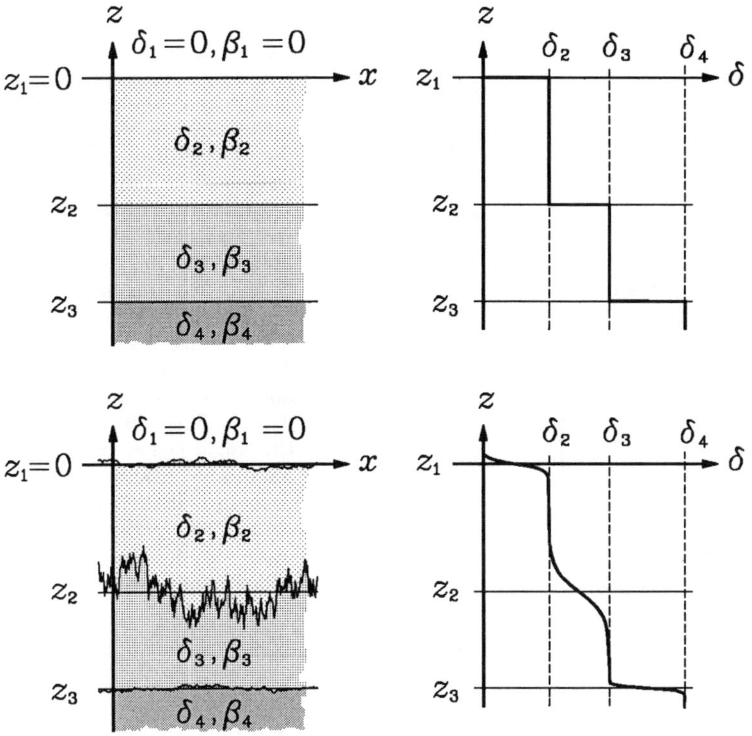

Fig. 2.16. System with sharp interfaces (*top*) and interfaces with small roughnesses (*bottom*). On the *left* an illustration of the interface structures and on the *right* the corresponding dispersion profiles $\delta(z)$ are shown

to be refined. After this "initial guess", the profile $\delta(z)$ is sliced into very thin layers of approximately 1 Å thickness[23] with uniform dispersions and sharp interfaces (see Fig. 2.18). For this system the reflectivity may be calculated using the *Parratt* formalism given by Eqs. (2.16) and (2.17). It is quite obvious that numerically this procedure is very time-consuming and the method based on the modified Fresnel coefficients $\tilde{r}_{j,j+1}$ should be applied whenever possible.

For systems possessing very thin layers, e.g. oxide layers, the condition $\sigma_j \ll d_j$ is quite often violated. However, a straightforward parametrization of the whole dispersion profile which is still based on the layer structure is indeed possible by introducing the so-called "effective-density model". This model provides a reasonable "initial guess" since it reduces to the independent-layer case for small roughnesses.

[23] The thickness of the slices depends on the system for which the reflectivity is calculated. It has turned out that 1 Å is a reasonable value for most materials in the x-ray regime (more details can be found in Ref. [206]).

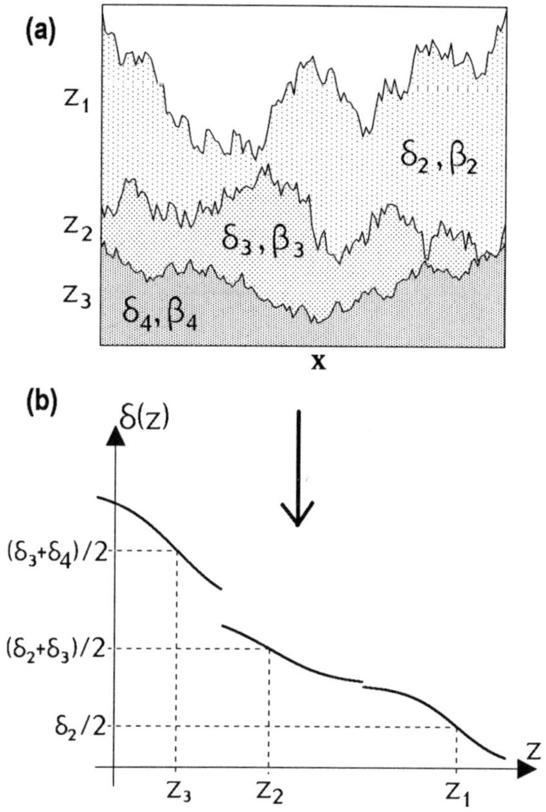

Fig. 2.17. (**a**) Layer system with large roughnesses. The term "large roughness" means that the layer thicknesses are on the same order of magnitude as the fluctuations at the interfaces, i.e. $d_j \approx \sigma_j$. The simple *Parratt* formalism is applicable only if the opposite condition $d_j \gg \sigma_j$ holds for *all* interfaces. (**b**) Dispersion profile obtained if the interfaces of the layer stack shown above were treated independently. The dispersions of the single layers are approximated by tanh functions with individual rms roughnesses. This parametrization leads to a model profile which is not continuous

The "effective-density model" is based on the following considerations: The profiles at the interfaces are essentially determined by functions $Y_j(z)$ with limits $Y_j(z) \to \pm 1$ for $z \to \pm\infty$, e.g. $Y_j(z) = \tanh[z\,\pi/(2\sqrt{3}\sigma_j)]$ or $\mathrm{erf}[z/(\sqrt{2}\sigma_j)]$ (see last section)[24]. Furthermore, the quantity

$$W_j(z) = \begin{cases} \frac{1}{2}\Big[1 + Y_j\,(z - z_j)\Big] & \text{for } z \le \zeta_j\,, \\ \frac{1}{2}\Big[1 - Y_j\,(z - z_{j-1})\Big] & \text{for } z > \zeta_j\,, \end{cases} \tag{2.62}$$

is the fraction of material j at position z. The coordinate

$$\zeta_j = \frac{\sigma_j z_{j-1} + \sigma_{j-1} z_j}{\sigma_j + \sigma_{j-1}} \tag{2.63}$$

denotes the depth at which the upper and lower profiles of interface j are connected continuously. In particular, for $j = N+1$ ($z_{N+1} = -\infty$, $\sigma_{N+1} = 0$,

[24] For clarity the discussion is restricted to odd functions $Y_j(z) = -Y_j(-z)$.

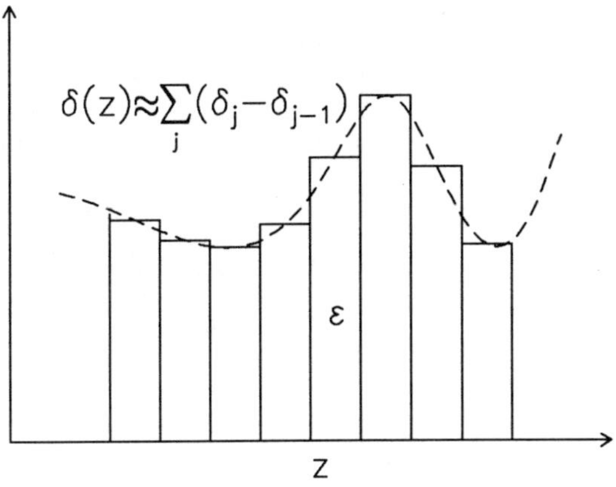

Fig. 2.18. A continuous dispersion profile $\delta(z)$ (*dashed line*) may be sliced into many very thin "layers" of thickness ε with constant δ_j values and sharp interfaces. Then $\delta(z) \approx \sum_j (\delta_j - \delta_{j-1})$ and the *Parratt* algorithm is applicable. Numerically this procedure is very time-consuming but nowadays quite simple to handle

substrate) one has $\zeta_{N+1} \to -\infty$, and $j = 1$ ($z_0 = +\infty$, $\sigma_0 = 0$, vacuum) leads to $\zeta_1 \to +\infty$. With Eq. (2.62) the dispersion profile $\delta(z)$ may be defined by

$$\delta(z) = \left(\sum_{j=1}^{N+1} \delta_j \, W_j(z) \right) \Big/ \left(\sum_{j=1}^{N+1} W_j(z) \right) . \tag{2.64}$$

Here the dispersion δ_j is the respective nominal value for material j. The parameters δ_j, σ_j, and $d_j = z_{j-1} - z_j$ are those which have to be refined. For small roughnesses, $\sigma_j \ll d_j$, the profile given by Eq. (2.64) splits into a system consisting of N independent layers. Therefore, calculating the reflectivity for this profile can be considered as a generalization of the procedure with the Fresnel coefficients $\tilde{r}_{j,j+1}$ for rough interfaces as outlined in the last section. However, the parameters do not have the same meaning as before. The "layer thickness" is similarly defined as $d_j = z_{j-1} - z_j$, which is the difference between the points of inflection of the $W_j(z)$ functions. The "roughness" σ_j determines the width of the intermediate region between material j and $j + 1$. If the interfaces are very rough this region may even be larger than the actual layer thickness. Then the dispersion is always much less than the nominal value δ_j within the interval $[z_{j-1}, z_j]$ and $\delta(z)$ given by Eq. (2.64) may be considered as the "effective dispersion at depth z", and hence, as the "effective density at depth z". Thus, this parametrization is referred to as the "effective-density model" [329, 354].

These rather theoretical considerations will now be explained by an example. Figure 2.19 shows a density profile for a layer system with large roughnesses. The profile was calculated for a PS film ($\delta_2 = 3.5 \times 10^{-6}$) of thickness $d_2 = 50$ Å on a silicon wafer ($\delta_4 = 7.56 \times 10^{-6}$) which is covered with an oxide layer of thickness $d_3 = 15$ Å ($\delta_3 = 6.8 \times 10^{-6}$). As the widths of the interfaces, the values $\sigma_1 = 8$ Å (vacuum/PS interface), $\sigma_2 = 13$ Å (PS/SiO_2 interface), and $\sigma_3 = 7$ Å (SiO_2/Si interface) were chosen. It is quite obvious that not all "layers" can be considered independently, as shown by the dashed lines

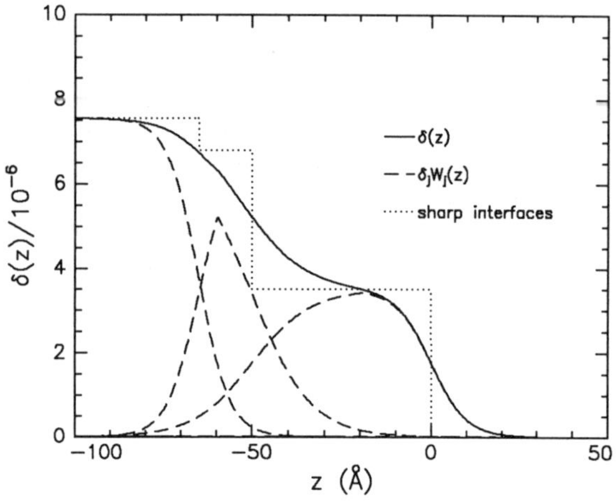

Fig. 2.19. Density profile of the layer system $Si/SiO_2/PS$ with large roughnesses. The *solid line* is the continuous dispersion profile $\delta(z)$ calculated according to the "effective-density model". The single "layers" are given by the dashed lines. For the thin oxide layer the nominal value of the dispersion is never reached. The *dotted line* shows the corresponding system with sharp interfaces

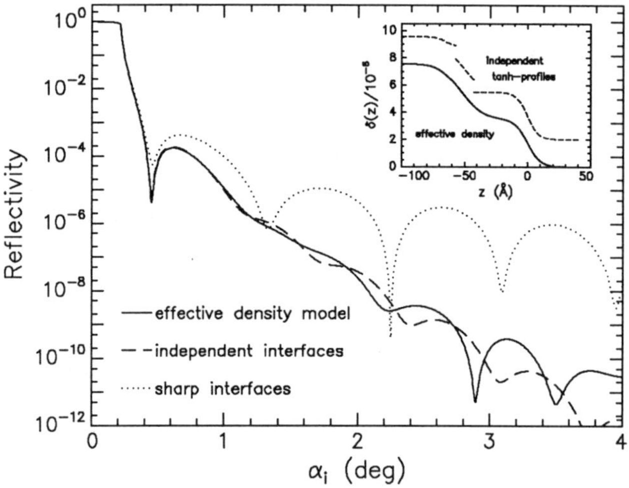

Fig. 2.20. Calculated x-ray (λ = 1.54 Å) reflectivities for a PS film of thickness d = 50 Å on Si/SiO_2. *Solid line*: effective density model. *Broken line*: Independent interfaces. *Dotted line*: without roughness. The dispersion profiles are depicted in the inset (shifted for clarity)

in Fig. 2.19. The dashed lines are the single contributions $\delta_j W_j(z)$ of each "layer". The consequence of roughness is most prominent for the very thin oxide "layer", which in fact just modifies the density profile of the rough Si surface.

The reflectivity calculated with the profile of Fig. 2.19 is shown in Fig. 2.20. The solid line indicates the effective density-model calculation and the dashed line is based on the assumption of independent layers with tanh interface profiles (*Parratt* algorithm with $\tilde{r}_{j,j+1}$ given by Eq. 2.38). The corresponding profiles are compared in the inset. It can be seen that the independent tanh model yields a dispersion profile with large discontinuities.

Whereas the reflectivities are almost identical for smaller incident angles, the differences become larger with increasing α_i. This means that, particularly for measurements which cover a large α_i or q_z range, the condition $\sigma_j \ll d_j$ has to be scrutinized carefully, and the independent-layer model – although nevertheless often applied – may not be valid for some systems.

The parametrization defined by Eq. (2.64) is one possibility for dealing with large roughnesses. The advantage is that single "layers" are regained by calculating the values $\delta_j W_j(z)$. However, in some cases other parametrizations are preferential as particular theoretical predictions exist[25]. Details will not be discussed here because this point will be addressed again in Chap. 4. In the next chapter several reflectivity experiments are presented which were quantitatively explained by the models given in the last sections.

[25] An example would be the phenomenon of critical adsorption. Here particular dispersion profiles are predicted theoretically (see e.g. *Flöter & Dietrich* [125], *Liu & Fisher* [222], or *Zhao et al.* [414], and references therein).

3. Reflectivity Experiments

This chapter is divided into two sections: Sect. 3.1 deals with setups for x-ray reflectivity experiments, and data analysis. Since nowadays reflectivity measurements are rather common, this section is more focused on the data analysis. Capillary waves with long-range correlations on soft-matter surfaces lead to subtleties in the data analysis (incorporation of the resolution) which are discussed in detail in Sect. 3.1.1. In the second part of this chapter, namely Sect. 3.2, x-ray reflectivity measurements from different soft-matter thin-film systems are presented.

3.1 Experimental Considerations

From the experimental point of view the most important challenges in surface x-ray scattering experiments are the glancing angles. This requires on the one hand very precise mechanics, which allow one to control the incident and exit angles with an accuracy of at least one-thousandth of a degree, and on the other hand very well-collimated radiation with low angular divergence. Nowadays the mechanics are no problem. Surface diffractometers are operating worldwide, where the precise motor movements are performed by computer-controlled systems. For x-ray sources the possibilities range from sealed tubes, rotating anodes, and synchrotron radiation facilities such as DORIS III/HASYLAB (Hamburg, Germany) or the NSLS (Brookhaven, USA), with bending-magnet, wiggler, or undulator beamlines, to so-called third-generation synchrotrons such as the ESRF (Grenoble, France), APS (Argonne, USA), and SPring-8 (Harima City, Japan).

Figures 3.1 and 3.2 show typical setups for standard reflectivity measurements. In the setup of Fig. 3.1 the radiation is generated by a rotating anode and collimated by a first slit before hitting the monochromator. Depending on the desired resolution and intensity, this may be a graphite crystal, a single crystal of silicon or germanium, or a channel-cut crystal, where the radiation is reflected two or more times to increase the resolution. After monochromatization the beam is again collimated by a second slit, before impinging onto the sample. The sample is mounted on a goniometer to control the incident

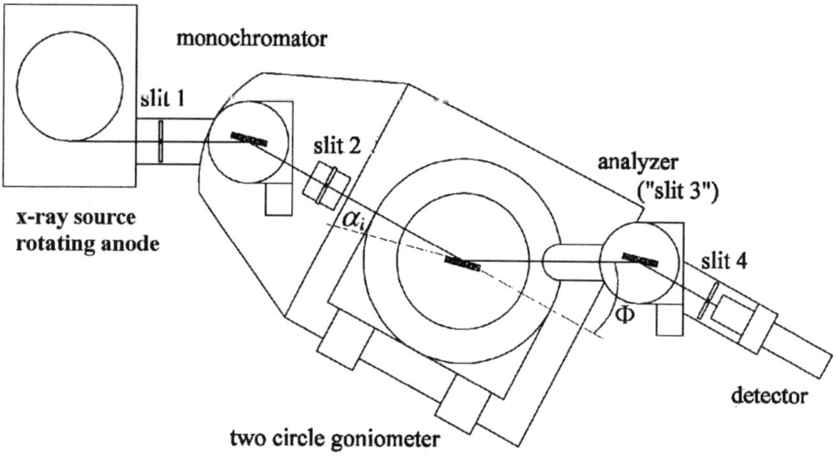

Fig. 3.1. Standard setup of a three-crystal diffractometer in the nondispersive (+ − +) configuration. The two slits at the incident side together with the monochromator define the angular divergence of the impinging radiation. The scattered x-rays are detected at an angle $\Phi = \alpha_i + \alpha_f$ with respect to the primary beam direction. The analyzer crystal ("slit 3") and the slit in front of the detector reduce the background and divergence of the outgoing beam

and exit angles, α_i and α_f, respectively[1]. At the detector side either a setup with an analyzer crystal or an arrangement with two slits in front of a detector unit is possible[2]. A typical synchrotron beamline setup is shown in Fig. 3.2, which, in principle, is identical to that of Fig. 3.1 (for details see *Feidenhans'l* [115]).

To investigate surfaces of bulk liquids special x-ray diffractometers have been constructed [7, 8, 9, 10, 12, 393]. The sample has always to be kept horizontal while a sophisticated monochromator setup bends the x-ray beam downwards onto the sample surface [47]. Since scattering from soft-matter *thin films* is mainly discussed here, gravity can be neglected, and this type of diffractometer is not required for most of the examples which will be presented later.

The wavevector transfer, $\boldsymbol{q} = \boldsymbol{k}_f - \boldsymbol{k}_i$, may be controlled by varying the incident and exit angles. The components are explicitly given by

$$q_x = k(\cos\alpha_f \cos\chi - \cos\alpha_i) , \tag{3.1}$$

$$q_y = k \cos\alpha_i \sin\chi , \tag{3.2}$$

$$q_z = k(\sin\alpha_i + \sin\alpha_f) , \tag{3.3}$$

[1] The angle $\Phi = \alpha_i + \alpha_f$ is often also denoted by 2Θ.

[2] In the regime of grazing angles the analyzer crystal simply acts as a very narrow slit, leading to smaller values of δ_{q_x} and δ_{q_z} (see below). In the vicinity of Bragg reflections the analyzer crystal changes the resolution function in a more complicated manner [52, 136, 224].

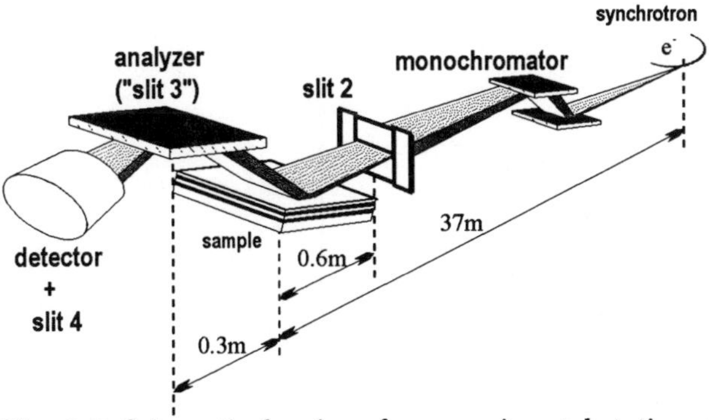

Fig. 3.2. Schematic drawing of an experimental station at a synchrotron beamline (here a bending-magnet beamline at HASYLAB) (figure taken from *Opitz* [269])

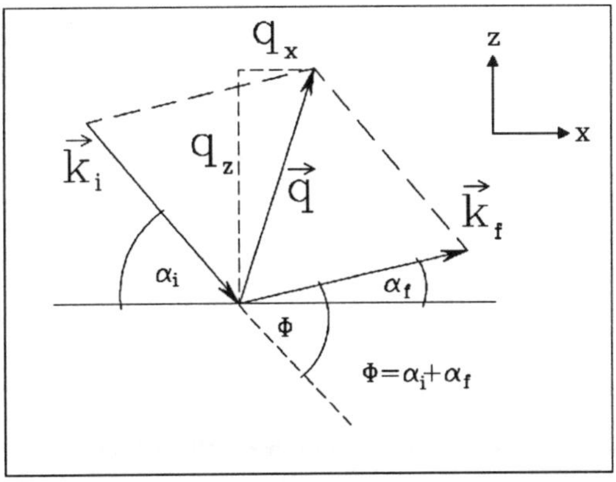

Fig. 3.3. In-plane scattering geometry: The wavevectors of the incident and scattered x-rays are $\boldsymbol{k}_i$ and $\boldsymbol{k}_f$, with the incidence and exit angles α_i and α_f, respectively, and the scattering angle Φ. The momentum transfer is defined by $\boldsymbol{q} = \boldsymbol{k}_f - \boldsymbol{k}_i = (q_x, q_z)$. For reflectivity $\alpha_i = \alpha_f$, corresponding to a q_z scan for $q_x = 0$

where the scattering plane is the (x, z) plane and the detector is placed at an angle χ out of this plane in the y direction. Reflectivity measurements ($\alpha_i = \alpha_f$) take place in the scattering plane, and thus $\chi = 0$ and $\boldsymbol{q} = (0, 0, 2k \sin \alpha_i)$. The in-plane scattering geometry is schematically shown in Figs. 3.3 and 3.4. In the case of GID experiments α_i and/or α_f are kept constant at small values, and the detector is rotated out-of-plane [12, 101, 102, 103, 167]. Also, diffuse scattering may be recorded in this geometry to obtain large wavevector transfers $q_r = (q_x^2 + q_y^2)^{1/2}$ parallel to the surface [248, 274, 275, 306, 308, 309, 310, 311, 312]. However, for soft-matter thin films long-range lateral correlations of capillary waves are of importance, so that high resolution in the scattering plane is required, while the resolution out-of-plane is totally relaxed, leading to an effective integration over q_y. We will come back to this point in Sect. 6.1.

X-ray scattering and particularly raw reflectivity data cannot be directly compared with model calculations for two reasons: (i) One has to take into account the finite resolution of the experimental setup. (ii) Often the data

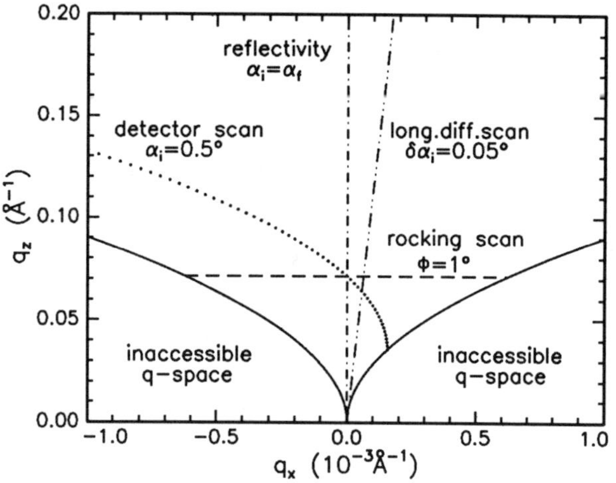

Fig. 3.4. Scans in reciprocal space (q_x, q_z). The region below the *solid line* is inaccessible for in-plane scattering (beam or detector below the sample). *Dashed line*: transverse scan. *Dashed–dotted line*: reflectivity. *Inclined dashed–dotted line*: longitudinal diffuse scan (reflectivity with offset $\delta\alpha_i$). *Dotted line*: detector scan

contain artefacts stemming from the scattering geometry, namely from the glancing angles. These topics will be discussed in the next two subsections.

3.1.1 Resolution Functions

A large body of work has been done concerning the calculation of resolution functions for various experimental configurations of x-ray diffractometers [51, 52, 61, 80, 136, 139, 224, 247, 260, 284, 410]. Here only the basic principles will be discussed. The resolution may be simply obtained from the total differentials of Eqs. (3.1) and (3.3), assuming grazing angles. In the case of in-plane scattering ($q_y = 0$) one obtains

$$\delta_{q_x} = \frac{\Delta\lambda}{\lambda} q_x + k(\alpha_i \Delta\alpha_i + \alpha_f \Delta\alpha_f) , \tag{3.4}$$

$$\delta_{q_z} = \frac{\Delta\lambda}{\lambda} q_z + k(\Delta\alpha_i + \Delta\alpha_f) , \tag{3.5}$$

where δ_{q_x} and δ_{q_z} are the resolutions parallel and perpendicular to the surface, the wavelength spread is $\Delta\lambda/\lambda$, and the angular divergence and acceptance are, $\Delta\alpha_i$ and $\Delta\alpha_f$, of the incoming and outgoing radiation, respectively. For most surface x-ray experiments $\Delta\lambda/\lambda < 10^{-4}$, and the monochromaticity of the radiation does not affect the resolution significantly[3]. If for the moment $\alpha_i \approx \alpha_f = \alpha$ and $\Delta\alpha_i \approx \Delta\alpha_f = \Delta\alpha$ are assumed, Eqs. (3.4) and (3.5) may be simplified to

[3] For neutron beamlines sometimes $\Delta\lambda/\lambda \sim 0.1$–$0.01$, with a remarkable effect.

$$\delta_{q_x} \approx q_z \, \Delta\alpha \quad \text{and} \quad \delta_{q_z} \approx 2k \, \Delta\alpha \, . \tag{3.6}$$

The angular divergence $\Delta\alpha_{\rm i}$ of the beam is mainly determined by the distances, slit sizes, and optical elements (mirrors, monochromator crystals) the radiation has to pass through on its way from the source to the sample. The acceptance of the detector unit $\Delta\alpha_{\rm f}$ is determined by the distances, slit sizes, and optical elements (analyzer crystals), too [80, 247, 269]. In fact, it may be quite complicated to calculate the divergence or resolution for a given setup from the specifications of all optical elements. However, it is easy to measure $\Delta\alpha$ in the small angle regime by determining the width of the primary beam without a sample[4], thus yielding via Eq. (3.6) good approximations for δ_{q_x} and δ_{q_z}. Typical values of resolutions used for reflectivity and diffuse-scattering experiments are $\delta_{q_x} \sim q_z \times 10^{-4}$ and $\delta_{q_z} \sim 10^{-3}\,\text{Å}^{-1}$ [52], which allow one to probe length scales laterally up to $x_{\rm max} \sim 2\pi/\delta_{q_x} \sim 10^5/q_z \sim 10^5$–$10^7\,\text{Å}$ [307, 369] and vertically up to $z_{\rm max} \sim 2\pi/\delta_{q_z} \sim 10^3$–$10^4\,\text{Å}$. The minimum length scales $x_{\rm min}$ and $z_{\rm min}$ accessible in an experiment are determined by the maximum wavevector transfer, which is restricted either by the rapid decrease of the measured intensity with increasing q_z or by the scattering geometry (see Fig. 3.4). Typical values are $x_{\rm min} \sim 100$–$1000\,\text{Å}$ for in-plane diffuse-scattering experiments and $z_{\rm min} \sim 5$–$10\,\text{Å}$, where one should note that rms roughnesses may be obtained from reflectivity data with an accuracy of approximately $\pm 0.1\,\text{Å}$. Out-of-plane diffuse-scattering measurements may probe lateral length scales considerably smaller than 100 Å (see the works of *Salditt et al.* [306, 309, 312]).

In the case of synchrotron experiments the divergence of the incident beam perpendicular to the electron orbit is given by $\Delta\alpha_{\rm i} \approx 1/\gamma = mc^2/E$, where γ is the Lorentz factor of the electrons with energy E. For $E \approx 5\,\text{GeV}$ one obtains $\Delta\alpha_{\rm i} \sim 0.1\,\text{mrad}$, which is often much smaller than $\Delta\alpha_{\rm f}$ and then our estimates above are not very accurate. This was investigated with a surface grating by *Gibaud et al.* [141], where a precise description of the calculation of the resolution δ_{q_x} from the slit settings for a particular synchrotron experiment is given.

In general, the resolution of an experiment in $\boldsymbol{q}$ space is given by a resolution function, $\tilde{R}(\boldsymbol{K})$, which determines both δ_{q_x} and δ_{q_z} [61, 80, 247]. Quite often a Gaussian[5]

$$\tilde{R}(\boldsymbol{K}) = \frac{1}{2\pi\delta_{q_x}\delta_{q_z}} \exp\Big\{ - K_x^2/(2\delta_{q_x}^2)\Big\} \exp\Big\{ - K_z^2/(2\delta_{q_z}^2)\Big\} \tag{3.7}$$

is used to model the resolution function. With this simple ansatz the q_x and q_z directions are treated independently and a time-consuming two-dimensional convolution of the respective scattering law (see below) with the resolution function may be avoided. A detailed investigation of this problem was recently

[4] Note that the FWHM $\Delta\Phi$ of the primary beam is $\Delta\Phi = 2\Delta\alpha$ since $\Phi = \alpha_{\rm i} + \alpha_{\rm f}$.

[5] Only in-plane scattering is considered here. The out-of-plane slits are assumed to be wide open, leading to an effective integration over q_y.

described by *de Jeu, Schindler & Mol* [80, 247]. It turns out that Eq. (3.7) is valid under the condition $\Delta\alpha_{\rm i} \approx \Delta\alpha_{\rm f}$, i.e. only if a symmetric setup is used for the experiment. Otherwise mixed terms such as $K_x K_z/(\delta_{q_x}\delta_{q_z})$ occur in the exponentials of Eq. (3.7) [52, 80, 247]. The measured intensity $I(\boldsymbol{q})$ in a detector placed at a point $\boldsymbol{q}$ in reciprocal space is connected with a theoretically given scattering law $\tilde{S}(\boldsymbol{K}) = |\tilde{\varrho}(\boldsymbol{K})|^2$ (e.g. the Fresnel reflectivity, Eq. 2.12) via a convolution

$$I(\boldsymbol{q}) = \int \tilde{S}(\boldsymbol{K})\, \tilde{R}(\boldsymbol{q} - \boldsymbol{K})\, \mathrm{d}\boldsymbol{K}\,. \tag{3.8}$$

In the case $\delta_{q_x}, \delta_{q_z} \to 0$, Eq. (3.7) yields $\tilde{R}(\boldsymbol{K}) \to \delta(\boldsymbol{K})$, thus leading to $I(\boldsymbol{q}) \to \tilde{S}(\boldsymbol{K})$ if the resolution is "infinitely good". Otherwise the finite resolution "smears out" the scattering law. This smearing out of the signal in reciprocal space is equivalent to an average over many regions of the sample in real space, since by the convolution theorem and Eq. (3.8) one may obtain [105, 340, 375, 377]

$$I(\boldsymbol{q}) = \iint \varrho(\boldsymbol{r})\varrho(\boldsymbol{r}')\, R(\boldsymbol{r}' - \boldsymbol{r}) \exp\left\{\mathrm{i}\,\boldsymbol{q}\cdot(\boldsymbol{r}' - \boldsymbol{r})\right\} \mathrm{d}\boldsymbol{r}\mathrm{d}\boldsymbol{r}'\,, \tag{3.9}$$

where $R(\boldsymbol{r})$ and $\varrho(\boldsymbol{r})$ are the Fourier transforms of $\tilde{R}(\boldsymbol{K})$ and $\tilde{\varrho}(\boldsymbol{K})$, respectively. The function $R(\boldsymbol{r})$ acts as a "cutoff function" in real space, and it may be considered as defining a "coherence volume" [340, 375, 377][6]. If this real-space cutoff is much smaller than the total sample volume, Eq. (3.9) is equivalent to the Fourier transform of the correlation function $\varrho(\boldsymbol{r})\varrho(\boldsymbol{r}')$ averaged on length scales equal to this cutoff over the entire sample.

On the surfaces of soft-matter films, capillary waves are often present (see Sect. 5.3.3) [53]. The spectrum of capillary waves contains long-range correlations on length scales comparable to or even larger than the above-mentioned coherence volumes. In fact, for capillary waves on bulk liquid surfaces, no separation between specularly and diffusely scattered intensity is possible. This complicates the analysis of x-ray reflectivity data since the measured roughness becomes resolution-dependent (see below).

Braslau et al. [47, 48] measured the surface roughness of water (see Fig. 2.5) and other liquids as a function of the experimental resolution. Similar experiments on alcohol surfaces were done by *Ocko et al.* [267] and *Sanyal et al.* [313]. Figure 3.5 depicts a reflectivity measurement of an ethanol surface. The symbols are the data and the line is a fit taking all resolution effects into account. It turns out that the curve may be well explained by the usual Fresnel reflectivity multiplied with a *Névot–Croce-* or *Beckmann–Spizzichino-*like roughness factor as explained in Sect. 2.3 (see also Sect. 6.1.3). However,

---

[6] With this approach the resolution is $\delta_{q_x} = 2\pi/L$, with L being a real-space length, namely the coherence length at the sample surface [103, 105]. Under what conditions this simple treatment is valid is the topic of Chap. 7.

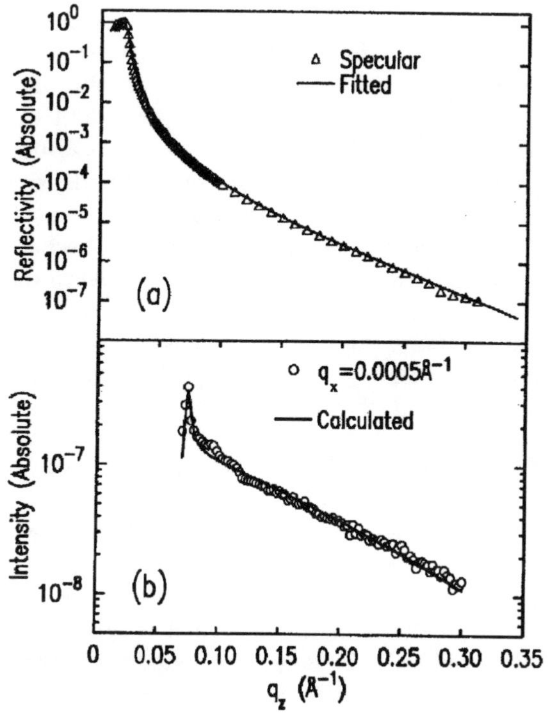

Fig. 3.5. **(a)** Measured (*symbols*) and fitted (*solid line*) "specular" reflectivity of a liquid ethanol surface at room temperature. The roughness (see Eq. 3.11) was $\sigma = (6.9 \pm 0.2)$ Å. Since long-range correlations due to thermally excited capillary waves cause the roughness the reflectivity essentially consists of diffuse scattering at $q_x = 0$. **(b)** Measured (*symbols*) and calculated (*solid line*) longitudinal diffuse scattering for an offset of $\Delta q_x = 0.0005$ Å^{-1} from the reflectivity condition. The intensities are shown on an absolute scale (figure taken from *Sanyal et al.* [313])

the roughness now is an "effective roughness" σ_{eff} which consists of three parts [313][7]:

$$\sigma_{\text{eff}}^2 = \sigma^2 + \frac{1}{2} B \gamma_{\text{E}} - \frac{1}{2} B \ln\left(\frac{\delta_{q_x}}{q_{\text{l,c}}}\right) , \tag{3.10}$$

where $\gamma_{\text{E}} = \lim_{m\to\infty} (\sum_{n=1}^{m} 1/n - \ln m) \approx 0.5772$ is Euler's constant [4], and[8]

$$\sigma^2 = \frac{1}{4} B \ln\left(\frac{q_{\text{u,c}}^2 + q_{\text{l,c}}^2}{q_{\text{l,c}}^2}\right) \tag{3.11}$$

is the total intrinsic mean-square surface displacement stemming from the capillary wave undulations[9]. In Eqs. (3.10) and (3.11) B is defined by

$$B = \frac{k_{\text{B}} T}{\pi \gamma} , \tag{3.12}$$

[7] For a deeper insight into the calculations leading to Eq. (3.10) see also *Dutta & Sinha* [105] and Sect. 6.1.3.

[8] For a calculation of this result from basic capillary wave properties see Sect. 5.3.3.

[9] It should be noted that a Gaussian probability density $P_j(z)$ (error-function profile $n(z)$, see Sect. 2.3) was assumed to obtain Eq. (3.10). For capillary waves this is a reasonable assumption [63].

where γ is the surface tension of the bulk liquid, T is the absolute temperature in kelvin, and $k_B = 1.3806 \times 10^{-23}$ J/K is Boltzmann's constant. At room temperature $B \sim 5$–$10\,\text{Å}^2$ for many liquids. The wavevector $q_{u,c}$ is an upper cutoff and may be set to $q_{u,c} = 2\pi/\kappa \sim 0.1$–$1\,\text{Å}^{-1}$ with κ the size of the molecules of the liquid (see *Ocko et al.* [267]). The quantity $q_{l,c}$ is a low-wavevector cutoff given by $q_{l,c} = \sqrt{\varrho g/\gamma} \sim 10^{-8}$–$10^{-7}\,\text{Å}^{-1}$ (ϱ is the density of the liquid and $g = 9.81\,\text{m/s}^2$ is the acceleration due to gravity). The cutoff $q_{l,c} = q_g = \sqrt{\varrho g/\gamma}$ is often referred to as the "gravitational cutoff" because it is determined by the ratio of the background gravitational potential to the surface tension.

The third term in Eq. (3.10) contains the resolution dependence of the effective roughness. Since the inequalities $q_{l,c} \ll \delta_{q_x} \ll q_{u,c}$ hold for bulk liquids, Eq. (3.10) can be simplified to [11, 267, 313]

$$\sigma_{\text{eff}}^2 = \sigma_{\text{in}}^2 + \frac{1}{2} B \ln\left(\frac{q_{u,c}}{\delta_{q_x}}\right) = \sigma_{\text{in}}^2 + \frac{1}{2} B \ln\left(\frac{2\,q_{u,c}}{q_z\,\Delta\alpha_f}\right), \tag{3.13}$$

where the result $\delta_{q_x} = (q_z/2)(\Delta\alpha_i + \Delta\alpha_f) \approx q_z\Delta\alpha_f/2$ for synchrotron experiments has been used, neglecting the small divergence $\Delta\alpha_i$ of the incident radiation compared to the angular acceptance $\Delta\alpha_f$ of the detector. Equation (3.13) explicitly shows the dependence of the measured roughness on the instrumental resolution[10]. It should also be noted that besides the thermal capillary wave term in Eq. (3.13) an additional intrinsic roughness σ_{in} accounting for the width of the liquid interface in the absence of capillary waves has been introduced. This intrinsic roughness is on the order of molecular dimensions and values of about $\sigma_{\text{in}} \sim 1$–$2\,\text{Å}$ are found in the literature [48, 267]. Often the intrinsic roughness is quite difficult to measure, since a separation of the capillary wave part from σ_{in} contains ambiguities. Recently, *Tostmann et al.* [379], however, were able to separate the capillary wave scattering of a liquid-metal surface from the contributions of the density profile, finally yielding the true intrinsic density profile at the liquid/vapour interface.

On liquid thin-film surfaces long-range correlations are suppressed by the strong substrate/film interaction. The propagation of capillary waves on bulk liquid surfaces is hindered only by gravitation (gravitational cutoff q_g, see above). The van der Waals interaction between a thin film of thickness d and a substrate is much stronger (see Sects. 3.2.1 and 5.3.3), hence eliminating all long-wavelength undulations. Therefore, the treatment of x-ray scattering data again becomes possible with the simple resolution-folding procedure given by Eq. (3.8). Thus, for thin liquid films one regains a narrow specularly reflected component of width δ_{q_x} riding on a broad diffusely scattered contribution, i.e. a clear separation of specularly and diffusely scattered radiation. Also, the surface roughness given by Eqs. (3.10) and (3.13) becomes

[10] It is also interesting to note that the measured effective roughness depends explicitly on the vertical wavevector transfer q_z.

resolution independent since for thin films $q_{\mathrm{l,c}} > \delta_{q_x}$ is always valid. This is discussed in more detail in Sects. 3.2.1 and 6.1.3.

3.1.2 Data Correction and Parameter Refinement

Besides the resolution function of a given experimental setup, several other corrections have to be taken into account. Either these corrections may be included in the theoretical formulas or the data may be corrected before a comparison with the theory. Factors caused mainly by the grazing angles are discussed first.

For very small incident angles the sample surface is almost parallel to the incident beam. Parts of the incoming radiation do not hit the sample, hence leading to a reflectivity less than unity in the total-external-reflection regime [139]. If a rectangular beam profile[11] of width b_i and a sample of length l are assumed, one may obtain from simple geometrical considerations that the angle of full illumination, α_b, is given by $\alpha_\mathrm{b} = \arcsin(b_\mathrm{i}/l)$. Since the specularly reflected intensity is proportional to the incident flux the following correction factor

$$\beta_{\mathrm{spec,i}} = \begin{cases} \sin\alpha_\mathrm{i}/\sin\alpha_\mathrm{b} & \alpha_\mathrm{i} \leq \alpha_\mathrm{b} \\ 1 & \alpha_\mathrm{i} > \alpha_\mathrm{b} \end{cases} \tag{3.14}$$

has to be taken into account (see Fig. 3.6). Whereas this factor always occurs in the interpretation of glancing-angle x-ray reflectivity data, a second factor, $\beta_{\mathrm{spec,f}}$, which accounts for the fact that the detector "looks" at a certain area of the sample, depends on whether or not one or two slits or analyzer crystals are used. If the slit sizes b_f at the detector side are larger than the incident-beam size b_i, then we simply have $\beta_{\mathrm{spec,f}} = 1$. Otherwise, the measured intensity is reduced by the constant factor $\beta_{\mathrm{spec,f}} = b_\mathrm{f}/b_\mathrm{i}$. Thus, the total geometrical correction for specularly reflected radiation is

$$\beta_{\mathrm{spec}} = \beta_{\mathrm{spec,i}}\,\beta_{\mathrm{spec,f}}\ . \tag{3.15}$$

Note that α_b, and hence β_{spec}, can be easily obtained during the alignment of a sample. A proper determination of this factor has to be done because the influence on the reflected intensity is quite large [139]. Figure 3.6 shows that this is of particular importance for small samples, i.e. for large illumination angles α_b. If $\alpha_\mathrm{b} > \alpha_\mathrm{c}$, then the decrease of the Fresnel reflectivity is also modified[12]. A wrong correction factor β_{spec} would have the same effect as a roughness, leading to incorrect results in a parameter-refining procedure.

In the case of diffusely scattered radiation the reverse situation is present. In Chap. 6 it will be explained that the diffuse-scattering cross section is proportional to the illuminated area of the sample. This area decreases with increasing incidence angle. Simple geometrical considerations lead to

[11] This is a crude approximation. All corrections may also be easily performed with Gaussian or Lorentzian beam profiles (see e.g. *Gibaud, Vignaud & Sinha* [139]).

[12] This is often the case. For $b_\mathrm{i} = 0.1\,\mathrm{mm}$ and $l = 10\,\mathrm{mm}$ one obtains $\alpha_\mathrm{b} = 0.57°$.

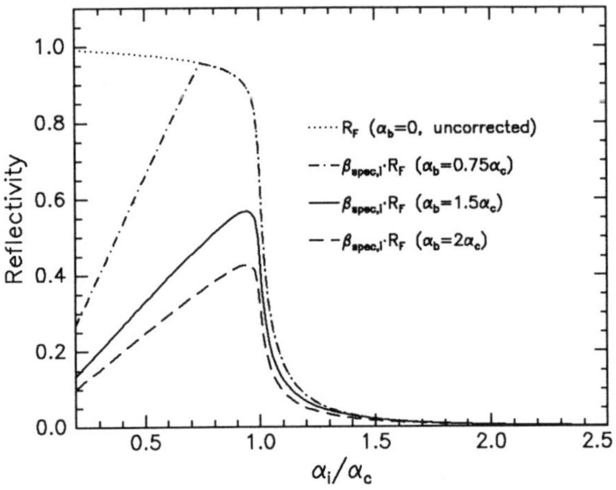

Fig. 3.6. Influence of the correction factor $\beta_{\mathrm{spec,i}}$ on the specularly reflected intensity for different illumination angles α_{b}. As parameters, $\delta = 7.56 \times 10^{-6}$ (silicon, $\alpha_{\mathrm{c}} = 0.22°$, $\beta/\delta = 1/50$) and an x-ray wavelength $\lambda = 1.54$ Å were used. Note, that the decrease of the Fresnel reflectivity is modified if $\alpha_{\mathrm{b}} > \alpha_{\mathrm{c}}$

$$\beta_{\mathrm{diff,i}} = \begin{cases} 1 & \alpha_{\mathrm{i}} \leq \alpha_{\mathrm{b}} \\ \sin\alpha_{\mathrm{b}}/\sin\alpha_{\mathrm{i}} & \alpha_{\mathrm{i}} > \alpha_{\mathrm{b}} \end{cases} \tag{3.16}$$

as the factor for the incident side and

$$\beta_{\mathrm{diff,f}} = \begin{cases} 1 & \alpha_{\mathrm{f}} \leq \arcsin(b_{\mathrm{f}}/l\beta_{\mathrm{diff,i}}) \\ b_{\mathrm{f}}/(l\,\beta_{\mathrm{diff,i}}\sin\alpha_{\mathrm{f}}) & \alpha_{\mathrm{f}} > \arcsin(b_{\mathrm{f}}/l\beta_{\mathrm{diff,i}}) \end{cases} \tag{3.17}$$

for the exit side. The total geometrical correction for diffusely scattered radiation is [122, 123, 354]

$$\beta_{\mathrm{diff}} = \beta_{\mathrm{diff,i}}\,\beta_{\mathrm{diff,f}}\ . \tag{3.18}$$

Furthermore, other corrections have to be taken into account. Specular-reflectivity data always consist of the reflected intensity plus the radiation diffusely scattered "by chance" in the direction $\alpha_{\mathrm{i}} = \alpha_{\mathrm{f}}$. Thus, the diffuse scattering has to be subtracted before comparing the data with the results of Chap. 2. This may be done by applying the condition $\alpha_{\mathrm{i}} + \delta\alpha_{\mathrm{i}} = \alpha_{\mathrm{f}}$, with a small offset $\delta\alpha_{\mathrm{i}}$, instead of $\alpha_{\mathrm{i}} = \alpha_{\mathrm{f}}$, while performing a scan. Here $\delta\alpha_{\mathrm{i}}$ has to be larger than the width of the specularly reflected beam. With such a "longitudinal diffuse scan" (see Figs. 3.4 and 3.5b) the q_z dependence of the diffuse scattering may be obtained and afterwards subtracted from the reflectivity data yielding the "true specular reflectivity". It should also be noted that other backgrounds may be subtracted similarly from the data when diffusely scattered intensities are compared with a theory. A procedure to separate bulk from surface scattering contributions for liquid surfaces is described by *Sanyal et al.* [313]. The particular background treatment used depends mainly on the detailed experimental conditions.

Since the dynamic range of x-ray reflectivity measurements covers more than six orders of magnitude, a parameter refinement is often done on a logarithmic scale[13]. This means that the function

$$\chi^2 = \sum_{j=1}^{N} g_j \left(\frac{\log R_{j,\text{exp}} - \log R_{j,\text{theo}}}{\Delta R_j} \right)^2 , \tag{3.19}$$

is minimized. The $R_{j,\text{exp}}$ are the background-corrected data points with statistical errors ΔR_j and weights g_j, and the $R_{j,\text{theo}}$ are calculated from a given theory including the resolution-folding and geometrical factors as explained above. The latter depend on parameters such as layer thicknesses d_j, roughnesses σ_j, etc., which have to be refined by least-squares-fitting algorithms. A nice review of different least-squares-minimizing routines and investigations of their convergence to a global minimum has been given by *Kosmol* [195]. Although nowadays sophisticated numerical algorithms and fast computers are available, it is still difficult to determine the precise error bars of the refined parameters. A way in which this may be achieved in a mathematically correct manner is described by *Jeß et al.* [181]. However, since the reflected intensity depends in a highly nonlinear and nontrivial manner on the density profile a simple error treatment is not possible.

At the end of this section it should be emphasized that in general x-ray reflectivity data cannot be inverted, i.e. it is not possible to obtain the density profile $\varrho(z)$ unambiguously from the measured data. Phase information is lost in calculating the reflectivity according to the formulas given in Chap. 2. This situation may be significantly improved by including a-priori knowledge of the system under consideration, together with special mathematical techniques. In Chap. 4 it will be shown how reflectivity data can be inverted to obtain the density profile directly without fitting. In most of the examples discussed in the next section the conventional parameter refinement, as described above, was applied.

3.2 Examples of Soft-Matter Thin Film Reflectivity

The presentation of the examples is divided into four subsections. It starts with the reflectivity of polymer films, where both thin films on silicon substrates and freestanding films are discussed. In general, polymer/polymer interfaces cannot be seen by x-ray reflectivity since the scattering contrast is too low. Here neutron reflectivity has one of its domains of application, because the scattering contrast can be enhanced by deuteration of one component. The interested reader is referred to the reviews of *Russell* [303, 304], *Stamm* [348], *Deutsch & Ocko* [85], and the conference proceedings edited

[13] Often also $q_z^4 R(q_z)$ or $R(q_z)/R_{\text{F}}(q_z)$ instead of $\log R(q_z)$ are refined. The rapid intensity decrease of the reflectivity $R(q_z)$ is compensated here by the factors q_z^4 or $1/R_{\text{F}}(q_z)$.

by *Felcher & Russell* [118] and references therein. Interesting further examples of neutron reflectivity from polymer/polymer interfaces are given in Refs. [150, 279, 325, 326, 332, 347, 374, 411]. Here the work of *Sferrazza et al.* [332] should be explicitly mentioned, in which the broadening of a polymer/polymer interface due to capillary waves was investigated for the first time. Capillary waves are the major topic of Sects. 3.2.1, 3.2.2, 5.3.3, 6.2.1, and 6.2.2.

The next two classes of systems are liquid thin films on solid and on liquid substrates, with particular emphasis on the former. Not all experimental details of preparing liquid homogeneous films are presented. The relevant information may be found in the quoted literature. Finally, the use of reflectivity measurements for determining the vertical structure of Langmuir–Blodgett films is demonstrated.

All of the examples were chosen because they show that x-ray reflectivity is an almost unique tool to extract structural information on an angstrom scale for many soft-matter films. They also demonstrate how powerful this technique is for various different samples with quite different physics involved. The examples range from simple measurements of the surface roughness and film thickness to precise determinations of interface structures and phase transitions. The reflectivities were measured up to large perpendicular wavevector transfers ($q_z \sim 1\,\text{Å}^{-1}$) with synchrotron radiation. Hence, atomic dimensions are accessible. However, it should be noted that the problem of radiation damage arises at modern synchrotron sources. This excludes several interesting systems from x-ray reflectivity investigations.

3.2.1 Polymer Films

The physics of polymers is an exciting and important field in materials science and basic research [100]. Bulk properties have been extensively investigated in the past and large tables such as the *Encyclopedia of Polymer Science and Engineering* were generated. The investigation of polymer/solid interfaces has great economic impact since multilayers made up of alternate metal and insulating layers play an increasing role in microelectronics (see e.g. Refs. [114, 138, 395] and references therein). Polymer thin films are used technologically as coatings, insulators, dielectric layers, and waveguides, and in many other fields. Additionally, they are of fundamental interest because the study of thermodynamic properties in confined geometries becomes possible. In very thin films the random coiling may be suppressed (see Fig. 3.7) and new phases with particular ordering may be found (for a recent example see *Sanyal et al.* [316]). In this section the main focus is on two fundamental properties of polymer films: (i) the influence of the constraint geometry on the surfaces roughness (ii) the possible difference between the glass transition temperature of a polymer thin film and the respective bulk value.

Before discussing the first point an application will be briefly explained, namely the application of polymer films as x-ray or neutron waveguides [345].

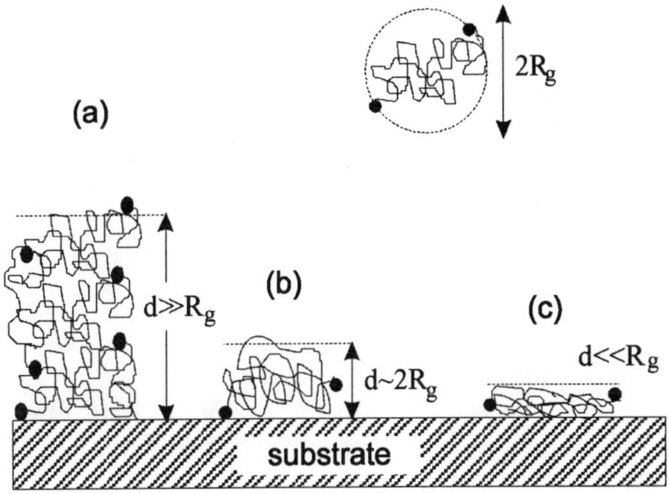

Fig. 3.7. Different confinements of polymer films on a solid substrate. The average end-to-end distance of a molecule is $2R_G$ (ends marked as bold circles). **(a)** $d \gg R_G$: no confinement. **(b)** $d \sim 2R_G$: intermediate state. **(c)** $d \ll R_G$: strong confinement

A polymer film of thickness $d \sim 1000$ Å serves as medium with low electron density. Above this medium a thin film with larger density is placed. Below the polymer is the substrate, for example a silicon wafer, which also has a larger electron density than the polymer (see e.g. *Feng et al.* [119]). Since the vertical structure of such an arrangement is equivalent to a potential with a pronounced minimum, x-rays are trapped in the polymer film by being totally reflected back and forth at the polymer/substrate and overlayer/polymer interfaces, respectively. It has been shown that a flux enhancement by a factor of 20 may be achieved by such a device [119, 120]. This guiding effect has also been observed for neutrons using similar materials [121]. Recently, the first such devices were used for x-ray phase contrast microscopy with submicron resolution by *Lagomarsino et al.* [179, 201].

The examples which will be discussed in more detail are thin films of polystyrene (PS) and polyethylene–propylene (PEP) with molecular weights $M_{W_{PS}} = 90$ k and $M_{W_{PEP}} = 290$ k, respectively (see Fig. 3.8). A quantity that is of fundamental importance for the properties of a polymer is the radius of gyration R_G [100]. This length describes the average size of a coil consisting of one polymer chain (see Figs. 3.7 and 3.8) and is a function of the molecular weight. For the two polymers investigated, the radii of gyration are $R_{G_{PS}} \approx 80$ Å and $R_{G_{PEP}} \approx 150$ Å. The evolution of the surface roughness as a function of the film thickness was investigated and compared to capillary wave calculations. Films of nominal thicknesses $d_{PS} = 30$–1800 Å in the case of PS and $d_{PEP} = 30$–500 Å for PEP were spin cast [327] onto silicon wafers which were previously etched by hydrofluoric acid. For the thinner films the

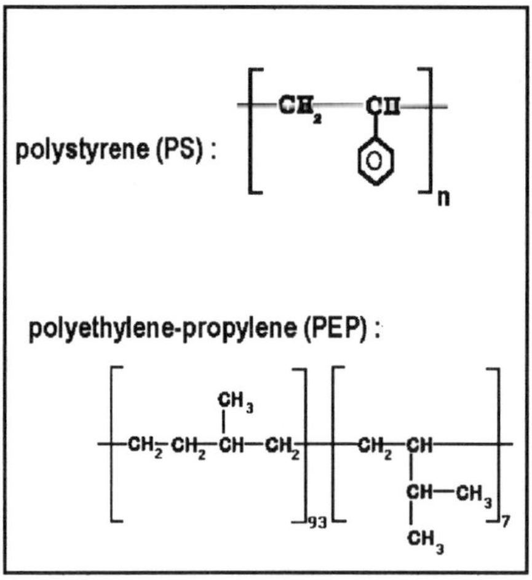

Fig. 3.8. Structures of the two polymers polystyrene (PS) and polyethylene–propylene (PEP). The glass transition temperature of PS is $T_{G,PS} \approx 100°C$ while that of PEP is $T_{G,PEP} \approx -62°C$. The structure of polyvinylpyridine (P2VP) is quite similar to that of PS. The only difference is that the second carbon atom of the benzene ring is replaced by a nitrogen atom. The radii of gyration R_G for the two polymers are given by: $R_{G_{PEP}} = 0.397\,M_{W_{PEP}}^{1/2}$ and $R_{G_{PS}} = 0.268\,M_{W_{PS}}^{1/2}$, where M_W is the molecular weight

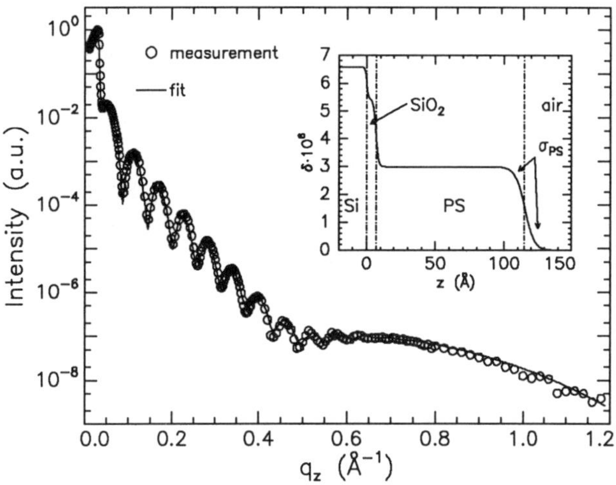

Fig. 3.9. X-ray reflectivity (*symbols*) and fit (*line*) of a $d = 109$ Å thick PS film deposited on an oxide-covered silicon wafer. The dispersion profile $\delta(z)$ shown in the inset was used for the refinement. The width σ_{PS} of the PS/air interface is of interest in this study [378]

confinement is most pronounced since the thicknesses are even smaller than the radius of gyration, hence hindering random coiling. The samples were annealed for approximately one day at 180°C under high-vacuum conditions ($p \sim 10^{-6}$ mbar) to maintain equilibrium conditions and to prevent dewetting [176, 177]. Then they were cooled down to room temperature and measured with synchrotron radiation (NSLS, beamlines X10 A/B with $\lambda = 1.19$ Å, and HASYLAB beamline E2 with $\lambda = 1.00$ Å).

When a glassy polymer is heated the chain mobility increases as the temperature increases. At a specific temperature, T_G, the polymer undergoes a phase transition from the glassy to a liquid-like state. Quantities such as the viscosity that sensitively depend on the mobility of the molecules

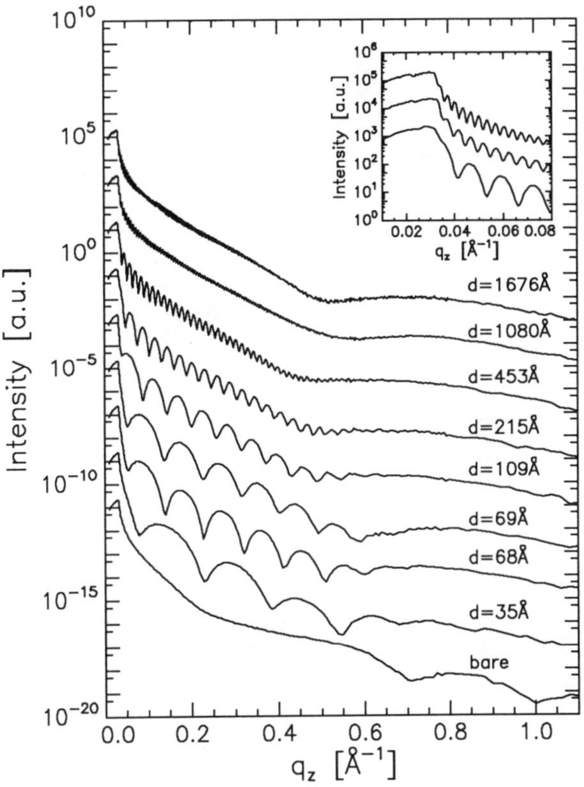

Fig. 3.10. X-ray reflectivities of polystyrene thin films for various film thicknesses *d*. The curve at the *bottom* is the reflectivity of the bare substrate. All curves were refined simultaneously with dispersion profiles quite similar to that shown in the inset of Fig. 3.9. For clarity the curves are shifted on the intensity scale and the fits are not shown. They are of the same quality as that given in Fig. 3.9. The inset displays a magnification of the region of the critical angle for the thickest three films [378]

change over several orders of magnitude. This temperature is called the glass transition temperature. For the two bulk materials $T_{G,PS} \approx 100°C$ and $T_{G,PEP} \approx -62°C$, are found. Therefore PEP is a liquid at room temperature whereas PS is quenched into a glassy state[14].

Figure 3.9 shows the reflectivity (symbols) of a $d = 109$ Å thick PS film on a silicon substrate as an example. The line is a fit performed with the dispersion profile shown in the inset. The substrate, a thin oxide layer, and the polymer film can be seen [378]. The density of the film was set to the nominal bulk value. The whole series of reflectivities is shown in Fig. 3.10 and was refined in the same manner. In particular, all substrate parameters were forced to be the same for the different films because all Si substrates were cut from the same wafer[15]. The overall shape of all curves is similar, whereas the modulation period decreases with increasing layer thickness. In

[14] As we will see later T_G may depend strongly on the film thickness so that the PEP films may also be quenched into a glassy state.

[15] This decreases the number of fit parameters. All Si/SiO_2 and SiO_2/PS interfaces were assumed to be identical to that displayed in the inset of Fig. 3.9 ($d_{SiO_2} = 15$ Å, $\sigma_{Si/SiO_2} = 4$ Å, $\sigma_{SiO_2/PS} = 3$ Å).

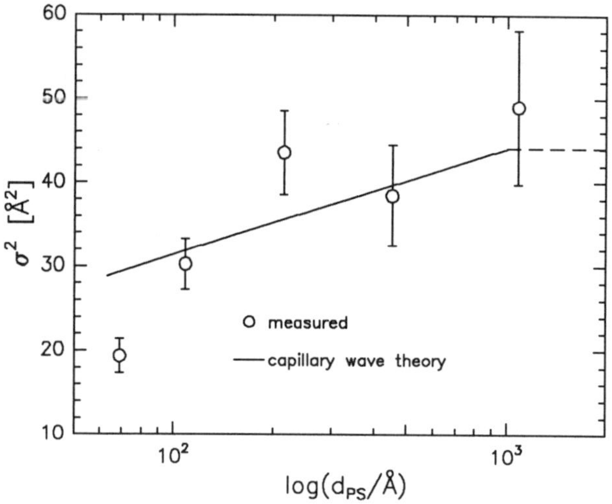

Fig. 3.11. Measured roughness σ^2_{PS} (*open circles*) vs. PS film thickness d. *Line*: Fit with the capillary wave model with van der Waals cutoff $q_{\mathrm{vdW}}(d) = a/d^2$. The thickest film already corresponds to a free film with $q_{\mathrm{vdW}}(d) < \delta_{q_x}$. *Broken line*: capillary wave roughness for a semi-infinite system [378]

the analysis the rms roughness σ_{PS} of the PS/air interface was of particular interest. The result for σ^2_{PS} versus the film thickness d is shown in Fig. 3.11.

A straightforward explanation of this curve would be the following: As already discussed in Sect. 3.1.1, for free capillary waves the tiny gravitational cutoff $q_{\mathrm{l,c}} = \sqrt{g\,\varrho/\gamma}$ is much smaller than the resolution δ_{q_x}. Hence $q_{\mathrm{l,c}}$ has been replaced by the q_x resolution, finally yielding Eq. (3.13). Since for the PS films a separation between the narrow specular component and the broad diffuse scattering was possible in the experiment[16] (see also Fig. 6.2 in Chap. 6), the lower cutoff must be larger than the resolution. If the capillary wave roughness given by Eqs. (3.11) and (3.13) is modified by replacing the lower cutoff $q_{\mathrm{l,c}}$ with[17]

$$q_{\mathrm{vdW}}(d) = \sqrt{\frac{A_{\mathrm{eff}}}{2\pi\gamma d^4}} = \frac{a}{d^2}\,, \tag{3.20}$$

one obtains

$$\sigma^2_{\mathrm{PS}}(d) = \sigma^2_{\mathrm{in}} + \frac{1}{2}B\ln\left[\frac{q_{\mathrm{u,c}}}{q_{\mathrm{vdW}}(d)}\right] \sim \ln d\,, \tag{3.21}$$

i.e. a logarithmic increase of the capillary wave roughness with the film thickness d for $q_{\mathrm{vdW}}(d) > \delta_{q_x}$ [128, 332]. Equation (3.20) defines the so-called van der Waals cutoff, which results from the much stronger van der Waals substrate/film interaction [106]. The effective Hamaker constant A_{eff} of the system Si/SiO_2/polymer describes the strength of this interaction [152, 387]. On the other hand the surface tension γ causes the film to be flat. The length

[16] More precisely, this separation was possible for $d < 1200$ Å, and hence not for the thickest film.

[17] A justification for this will be given in Sect. 5.3.3.

$a = \sqrt{A_{\text{eff}}/(2\pi\gamma)}$ is the ratio of these two quantities with competing effects. For most liquids and polymers this length is on the order of $a \sim 5$–10 Å [387].

It should be noted that the roughness has to saturate at the nominal free capillary wave value for very thick films because $q_{\text{vdW}}(d) \to 0$ for $d \to \infty$, thus regaining Eqs. (3.11) and (3.13). A more subtle calculation, where the integral over the power spectral density (see Chap. 5 and Sect. 5.3.3) is evaluated, reveals [378]

$$\sigma_{\text{PS}}^2(d) = \sigma_{\text{in}}^2 + \frac{1}{4} B \ln \left[\frac{q_{\text{u,c}}^2 + q_{\text{vdW}}^2(d)}{\delta_{q_\parallel}^2 + q_{\text{vdW}}^2(d)} \right] , \tag{3.22}$$

with the correct asymptotic behavior for large film thicknesses. Here $\delta_{q_\parallel}$ is the lateral resolution, which is assumed to be isotropic. If the more realistic situation of a narrow q_x resolution and an effective integration over q_y, as described in Sect. 3.1.1, is considered, the slightly different result

$$\sigma_{\text{PS}}^2(d) = \sigma_{\text{in}}^2 + \frac{1}{2} B \ln \left[\frac{q_{\text{u,c}} + \sqrt{q_{\text{u,c}}^2 + q_{\text{vdW}}^2(d)}}{\delta_{q_x} + \sqrt{\delta_{q_x}^2 + q_{\text{vdW}}^2(d)}} \right] \tag{3.23}$$

is obtained. Numerically, Eq. (3.23) is almost indistinguishable from Eq. (3.22).

The line in Fig. 3.11 is a fit of Eq. (3.23) to the data points with $a = (10 \pm 2)$ Å, $B = (12 \pm 2)$ Å^2, and the intrinsic roughness $\sigma_{\text{in}} = 1$ Å [267]. The fit is quite insensitive to the upper cutoff, which was set to $q_{\text{u,c}} = 1$ Å^{-1}, corresponding to the dimensions of the monomers. However, these values would suggest that the surface tension γ is lowered by a factor of two if A_{eff} is assumed to be correct. It is not quite clear yet whether the surface tension is affected in thin films by the constrained geometry. This question will be addressed again in the next subsection.

Although the errors in Fig. 3.11 are rather large the logarithmic increase of the surface roughness with the film thickness is visible. However, this behavior cannot serve as proof of the presence of capillary waves on a thin PS film surface. Here the diffuse scattering reveals more details. It shows that (i) the intrinsic roughness does not stem from roughness propagation from the underlying substrates[18], and that (ii) the ansatz with a thickness-dependent cutoff is justified (see Sects. 5.3.3 and 6.2.1) but a simple van der Waals cutoff cannot explain all features of the scattering data. Hence a modified film thickness dependence of $q_{\text{l,c}}$ has to be assumed, and the more general form

$$q_{\text{l,c}}(d) = a/d^m \tag{3.24}$$

is chosen, with $m = 2$ for pure van der Waals substrate/adsorbate interactions (see Eq. 3.20). If $m = 1$ instead of $m = 2$ then Eq. (3.21) suggests that this corresponds to a surface tension lowered by a factor of two as observed

[18] No conformal roughness was observed even for the thinnest films with $d < R_G$.

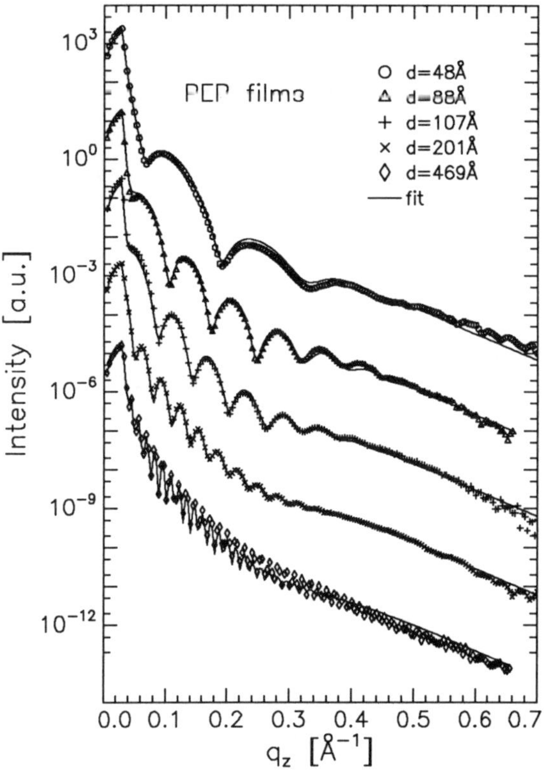

Fig. 3.12. Specular reflectivities (*symbols*) of PEP thin films for various film thicknesses and the respective fit results (*lines*). All curves were refined simultaneously with the layer thickness d and PEP/air interface widths σ_{PEP} as free parameters. The observed evolution of the surface roughness $\sigma_{\mathrm{PEP}}(d)$ is shown in Fig. 3.13, where $\sigma_{\mathrm{PEP}}(d)$ is the microscopic part of the roughness which is superimposed over the island substructure (see text and Fig. 3.14b). A dispersion profile similar to that given in the inset of Fig. 3.9 was assumed. All curves are shifted on the intensity scale for clarity [376]

above. As will be discussed in Sect. 6.2.1, the diffuse scattering also suggests a dependence $q_{\mathrm{l,c}} = a/d$ rather than the pure van der Waals type $q_{\mathrm{l,c}} = a/d^2$. This can be explained quantitatively by the theory of *Fredrickson et al.* [128] (see Sect. 5.3.3 and Eqs. 5.38–5.41), where it is essentially assumed that the strong polymer/substrate interaction transforms the quasi-liquid polymer film into a very viscous gel.

A similar series of measurements of polyvinylpyridine (P2VP) thin films on silicon substrates supports this picture. P2VP sticks even more strongly than PS to the SiO_2 surface, thus drastically changing the polymer/SiO_2 interaction. However, the detailed data analysis reveals that this has no drastic effect on the rms roughness of the P2VP/air interface [378]. But the analysis of the diffuse scattering now yields $m = 0.8$, in very good agreement with the the value $m = 0.75$ given by the above-mentioned theory of *Fredrickson et al.* [128] (see Sect. 6.2.1).

The results for the PEP films are quite different. Figure 3.12 depicts the reflectivities and fit results for the PEP series. The data look quite similar to

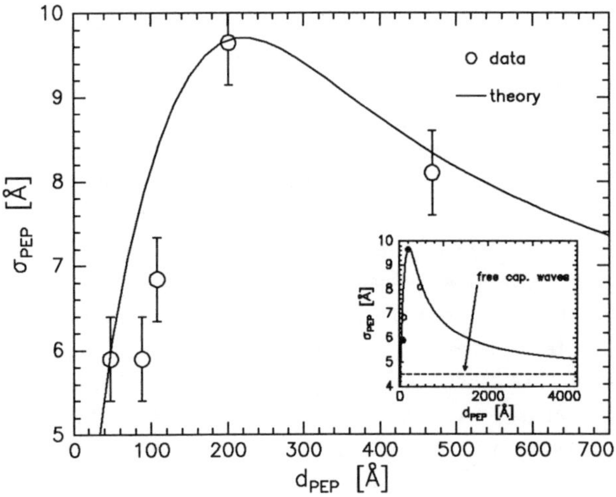

Fig. 3.13. Roughness σ_{PEP} (*open circles*) vs. PEP film thickness d_{PEP}. The *line* is a fit with the model described in the text, which is essentially based on a partial dewetting of the PEP films. The inset shows the calculated curve on a larger scale with the limit of free capillary waves (*dashed line*) reached for very thick films [376]

those shown in Fig. 3.10. But the refinement[19] reveals that the rms roughness σ_{PEP} of the PEP/air interface shows a maximum for film thicknesses of about $d_{PEP} \approx 200\,\text{Å} \sim R_G$. Figure 3.13 depicts σ_{PEP} as a function of d_{PEP} for the five investigated films. Obviously a simple van der Waals cutoff leading to a logarithmic increase of σ_{PEP} cannot explain the data.

The difference between the roughness curve $\sigma_{PS}(d)$ in Fig. 3.11 and $\sigma_{PEP}(d)$ shown in Fig. 3.13 is caused by the different wetting behaviors of PS and PEP on Si surfaces [270, 295, 413]. The spreading parameter $S_X = \gamma_{SiO_2} - (\gamma_{X,SiO_2} + \gamma_X)$, where X = PS, PEP and the interfacial tensions of the polymer/substrate, air/substrate, and polymer/air interfaces are γ_{X,SiO_2}, γ_{SiO_2}, and γ_X), respectively, describes the wettability of a SiO_2 surface with an adsorbate X [16, 76, 91, 161, 205]. If $S_X > 0$, then the surface wets completely, i.e. a homogeneous layer is stable. Otherwise the material X dewets the substrate and droplets are formed with a finite contact angle $\Theta = (\gamma_{SiO_2} - \gamma_{X,SiO_2})/\gamma_X$ [408]. Whereas $S_{PS} > 0$ is valid in any cases, and hence PS wets an oxide-covered surface[20], the situation for PEP is more complex. Here the spreading parameter is $S_{PEP} \approx 0$ and the wetting/dewetting behavior is quite sensitive with respect to the sample preparation. In particular, experiments of *Zhao et al.* [413] reveal that thin PEP films ($d < R_G$) dewet a silicon surface with a native oxide layer. The diameter of the PEP droplets or islands is an increasing function of the film thickness. Thicker films ($d \gg R_G$) wet the surface completely. On the other hand an oxide-free

[19] This was done in the same manner as for PS, with a very thin oxide layer of thickness $d_{SiO_2} = 10\,\text{Å}$. The same profiles at the Si/SiO_2 and SiO_2/PEP interfaces ($\sigma_{Si/SiO_2} = 4\,\text{Å}$, $\sigma_{SiO_2/PEP} = 3\,\text{Å}$) were assumed for all films.

[20] If thin PS films are not annealed under high-vacuum conditions as described above, they nevertheless dewet a SiO_2 surface [176, 177, 270].

Si surface would be completely wetted by PEP in any case. The etching process removes the native oxide from the silicon wafer. However, since this was done in air a new but thinner oxide builds up before the films are spun on the substrate. AFM measurements of all samples have revealed that the PEP films wet these surfaces coated with very thin "fresh" oxides. Obviously the wetting behavior of PEP on SiO_2 covered silicon wafers is a function of both the PEP and the SiO_2 film thicknesses. But the AFM images have shown more details: A scenario as sketched in Fig. 3.14b is found. The PEP films on the "fresh" oxides dewet on a small vertical length scale (see Fig. 6.8 in Sect. 6.2.1). They show an island substructure with an island height h on the order of $h \sim 30$–50 Å. Their lateral size $l(d)$ increases with the PEP film thickness and the island substructure vanishes for very thick films. This indicates that the spreading parameter S_{PEP} is close to zero for the investigated samples and these films are in the initial capillary instability stages of dewetting where the sample surface is just roughened on a mesoscopic scale (islands) but a complete dewetting down to the substrate has not yet occurred. The theory of *Brochard-Wyart et al.* [49] for resonant capillary modes which are present at the pre-rupture stage of an unstable film in the initial stage of dewetting yields a power-law behavior for the respective resonant wavenumber. If we assume that our dewetted structures have resulted from the evolution of these amplified fluctuations, the wavenumber $q_0 = 2\pi/l(d)$ may be estimated from $q_0 \propto d^{-3/2}$ [49].

The microstructure at the PEP surfaces is superimposed on the capillary wave roughness with a second thickness-dependent wavenumber q_0. If this is taken into account for the respective power spectral density (see Sects. 5.3.3 and 6.2.1) one obtains [376]

$$\sigma^2_{\mathrm{PEP}}(d) \approx \sigma^2_{\mathrm{PS}}(d) + \frac{B}{2\,d\,q_{\mathrm{vdW}}(d)} \arctan\left[q_{\mathrm{u,c}}/q_{\mathrm{vdW}}(d)\right], \tag{3.25}$$

where $\sigma^2_{\mathrm{PS}}(d)$ is the result given by Eqs. (3.22) or (3.23) with $q_{\mathrm{vdW}} = a/d^2$. We have set $q_0 \sim 1/d$, which is a good approximation for the island wavenumber as deduced from AFM images. However, it has to be mentioned that the theoretically predicted relation $q_0 \propto 1/d^{3/2}$ is still within the experimental errors. The line in Fig. 3.13 is a fit of the data using Eq. (3.25) with $B = (6 \pm 1)$ Å^2, $a = (8 \pm 2)$ Å, $q_{\mathrm{u,c}} \approx 0.05$Å^{-1}, and $\sigma_{\mathrm{in}} = 2$ Å. The parameters B and a agree within the errors with the values calculated from the Hamaker constant and the surface tension of PEP. Also, the value of the intrinsic roughness is reasonable. It is interesting to note that the data can be explained with a simple van der Waals substrate/adsorbate interaction. Since PEP does not stick to a silicon surface this is in accordance with the above-mentioned findings for PS and P2VP films. Without substrate/adsorbate interaction the theory of *Fredrickson et al.* [128] is not applicable, and PEP films should behave like simple van der Waals liquids. .

As already mentioned for the PS films, the explanation of the specular reflectivity only gives a hint of the exact surface morphology. In Sect. 6.2.1

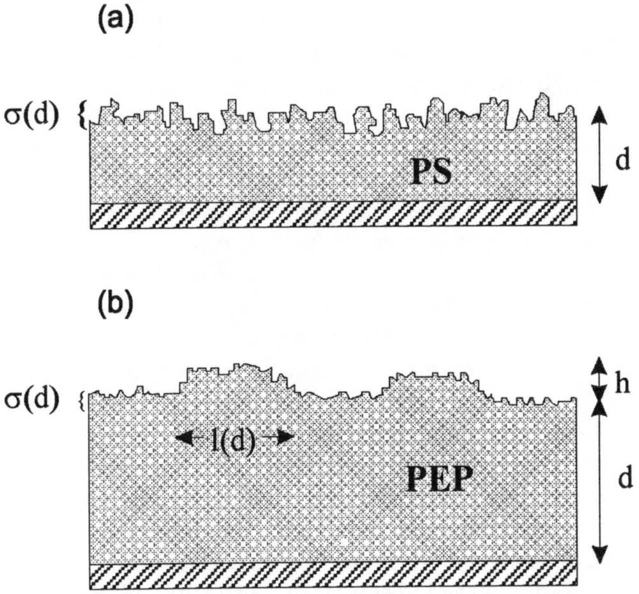

Fig. 3.14. Different surface morphologies of thin PS and PEP films: **(a)** PS films: waves with a thickness-dependent cutoff leading to a logarithmic increase of the surface roughness $\sigma(d)$ with the film thickness. **(b)** PEP films: initial stage of dewetting with an island substructure which is superimposed over the capillary wave fluctuations that lead to the microscopic roughness $\sigma(d)$. An additional wavenumber $q_0 = 2\pi/l(d)$ is introduced by the island structure (see also Fig. 6.8)

it will be shown that the explanation of the diffuse scattering from the films confirms almost all assumptions and conclusions.

The above examples describe systematic studies of soft-matter surfaces. The thickness dependence of the cutoff introduces the polymer/substrate interaction into the $\sigma(d)$ curves. Even for the strongly confined PEP films with $d < R_G$, capillary waves seem to be present at the surfaces. However, there are still open questions: The question of whether polymer surfaces are altered while passing through the glass transition is open[21]. Prelimary results indicate that the surface roughnesses of PS films in the liquid and the glassy state are identical [20, 331]. The glass transition is also the major topic of the next example of polymer thin films.

Another very interesting class of systems is freestanding soft-matter thin films. Here x-ray scattering is also an almost unique probe to obtain structural information since most other techniques are not able to cover the required q range. Further, these films have to be measured in rather complex experimen-

[21] This means only their static average. The dynamics above and below the glass transition temperature are quite different.

tal environments. *Gierlotka et al.* [142, 202] reported in the year 1990 an x-ray reflectivity study of fluctuations in thin freestanding smectic films. These investigations were later continued and completed by diffuse-scattering studies [81, 246, 247, 248, 334] (see Sect. 6.2.1). In 1992, *Daillant & Bélorgey* [68, 69] presented x-ray scattering experiments from ultrathin soap films ($d \sim 100$–400 Å, so-called "Newton black films"). Their experiments and calculations reveal that film parameters such as the surface tension, bending rigidity, and compression moduli may be extracted from the x-ray data.

For polymer thin films an interesting question is whether the glass transition temperature T_G is a function of the film thickness. However, the measurement of T_G in thin films is quite difficult. Hence, it is not surprising that different statements can be found in the literature. Whereas *Keddie, Jones & Cory* [184] report results obtained with ellipsometry which suggest that T_G decreases with decreasing film thickness, *Wu, Wallace & van Zanten* [383, 390, 399] describe the opposite effect: They found an increase of T_G with decreasing thickness from x-ray reflectivity data. The glass transition temperature was identified as the temperature where the thermal expansion coefficient $\alpha_{\text{th.exp.}}$ underwent an abrupt change [398]. The two studies were done on PS films spun cast onto etched Si substrates. The only difference was that in the former the PS films were heated in air, while in the latter study this was done in vacuum ($p \sim 10^{-4}$ Pa). However, this small difference may have a drastic influence on the wetting behavior [176, 177].

Keddie, Jones & Cory [184] explained their finding of a reduced glass transition temperature by the existence of a "liquid-like" layer near the free surface of the film. This liquid surface layer increases in thickness with increasing temperature. An influence of the substrate on T_G was not considered. Subsequent ellipsometry measurements by the same group revealed that for PMMA films on oxide covered Si substrates T_G increases with decreasing film thickness, while on gold substrates the opposite effect was found [185]. Thus, a strong influence of the substrate may not be ruled out.

Recent positronium annihilation investigations by *DeMaggio et al.* [82, 404] and *Jean et al.* [180] also suggest that the glass transition temperature of PS films decreases with decreasing film thickness. However, the results of *DeMaggio et al.* are more complicated. Their data were explained by assuming a constrained layer of thickness 50 Å at the Si interface and a surface region of thickness 20 Å with reduced T_G.

Theoretically the situation is as follows: Results from molecular dynamics simulations and calculations predict that the effect of a free surface is to decrease the density and enhance the mobility of the polymer chains near a polymer/vacuum interface [234, 238]. On the other hand Monte Carlo simulations by *Baschnagel & Binder* [26] show that a confining wall results in a density increase and therefore in a decrease of the chain mobility. Thus, theoretically it is suggested that the effect of a free polymer surface is to lower T_G while that of a substrate is to increase T_G. Qualitatively one may

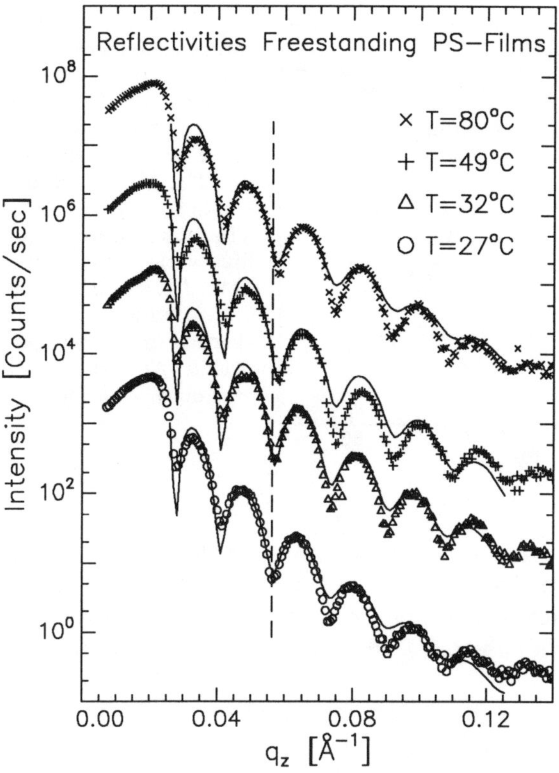

Fig. 3.15. Series of reflectivities (*symbols*) for a freestanding PS film of initial thickness $d_0 = 360\,\text{Å}$. The *solid lines* are calculations for a one-layer model. In the region of very low q_z (total external reflection regime) the data were not taken into account in the analysis. The films were slightly bent and hence a reliable roughness analysis was not possible. The *vertical dashed line* demonstrates that the period of the oscillations increases slightly with increasing film temperature. The curves are shifted on the intensity scale with respect to each other for clarity [19]

conclude that the result of an experiment depends on which of these effects dominates.

The asymmetric situation for a thin film on a substrate suggests that one should carry out measurements on freestanding polymer films. This was recently done by *Forrest et al.* [126] for freestanding PS films. Using Brillouin light scattering they found a linear decrease of T_G with decreasing film thickness. Since no raw data are presented it is difficult to decide how sensitive the method actually is. Furthermore, the measurements were not carried out in vacuum (see above).

Thus, x-ray reflectivity investigations performed under vacuum conditions on freestanding PS films have been carried out recently [19, 20]. Two series of samples will be discussed: (i) Three freestanding PS ($M_W = 660\,\text{k}$, $R_G \approx 200\,\text{Å}$, $T_{G_{\text{bulk}}} \approx 100°\text{C}$) films of initial film thicknesses $d_0 = 175\,\text{Å} \approx R_G$, $d_0 = 360\,\text{Å} > R_G$, and $d_0 = 992\,\text{Å} \gg R_G$. The films were spin cast onto glass substrates and then floated onto razor blades with an almost rectangular hole of size $3\,\text{mm} \times 6\,\text{mm}$ in the middle. Afterwards the films were annealed for several hours at $T \sim 90°\text{C}$ to achieve equilibrium conditions. (ii) A control series of three PS ($M_W = 90\text{k}$, $R_G \approx 80\,\text{Å}$, $T_{G_{\text{bulk}}} \approx 100°\text{C}$) films of initial

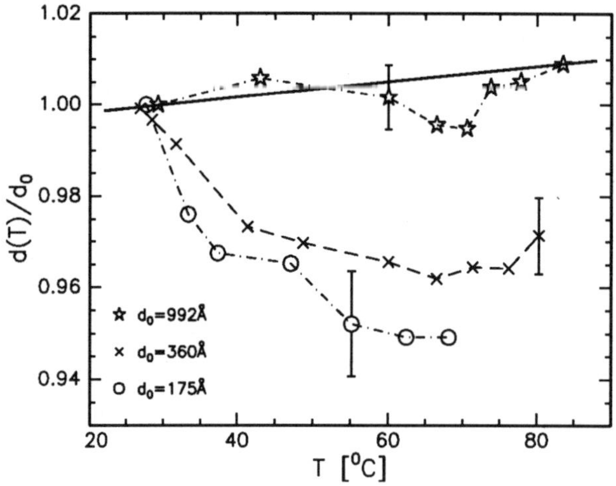

Fig. 3.16. Temperature-dependent relative thickness changes $d(T)/d_0$ of freestanding PS films as measured by x-ray reflectivity. The films were unstable above $T \sim 90°C$. The *solid line* for the film with initial thickness $d_0 = 992$ Å was calculated with the bulk thermal expansion coefficient. All other lines are guides to the eye [19]

film thicknesses $d_0 = 45$ Å $< R_G$, $d_0 = 203$ Å $> R_G$, and $d_0 = 863$ Å $\gg R_G$, which were spin cast onto etched silicon wafers and annealed as described before. Afterwards the surfaces were examined by AFM to check whether the PS had completely wetted the substrates, which is particularly important for the very thin films.

Since on the one hand the films were measured in situ under high-vacuum conditions, and on the other hand long annealing times were required to achieve thermal equilibrium during the measurements, the data were collected using a laboratory x-ray source (Cu$K\alpha_1$ radiation of wavelength $\lambda = 1.54056$ Å).

As example Fig. 3.15 shows a series of reflectivities of the $d_0 = 360$ Å freestanding PS film for different temperatures. Only the region $0.025\,Å^{-1} < q_z < 0.13\,Å^{-1}$ was used for the analysis (solid lines in Fig. 3.15). The data for $q_z < 0.025\,Å^{-1}$ (total-external-reflection regime) have to be excluded since the film was slightly bent over the hole in the razor blade. The reflected signal is partly suppressed in this region. In the region of large q_z the intensity decreases rapidly until the reflected signal vanishes for $q_z > 0.14\,Å^{-1}$. However, the oscillations within the measured q_z region allow one to obtain the layer thicknesses accurately. Transverse scans reveal that the specular peak is broadened. This indicates the influence of the bent edges of the film. Therefore, only crude estimates for the rms roughnesses of the interfaces can be given.

The relative thickness changes $d(T)/d_0$ as a function of the temperature T are plotted in Fig. 3.16. It can be seen that the thinner films with initial thicknesses $d_0 = 175$ Å and $d_0 = 360$ Å contract significantly (about 3–5%) with increasing temperature, whereas the thickness of the film with $d_0 = 992$ Å remains almost constant. This may be explained by the small thermal

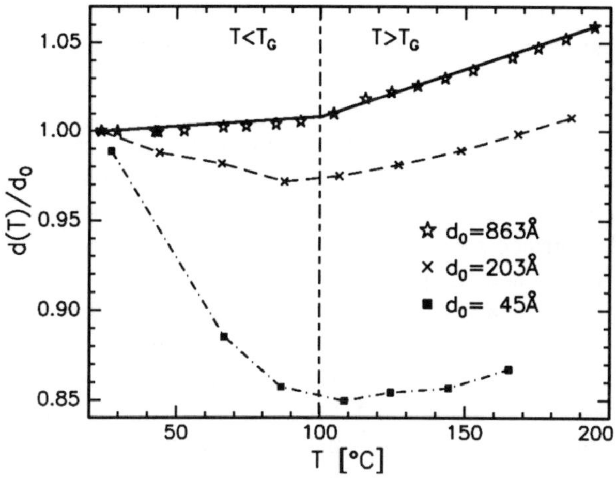

Fig. 3.17. Temperature-dependent relative thickness changes $d(T)/d_0$ of PS films on silicon substrates as measured by x-ray reflectivity. The *line* for the film with the initial thickness $d_0 = 863$ Å is calculated with the bulk thermal expansion coefficients of PS. The other lines are guides to the eye [19]

expansion coefficient $\alpha_{\text{th.exp.}} \approx 1.7$–$2.1 \times 10^{-4}\,\text{K}^{-1}$ of bulk PS below T_G [46] (solid line in Fig. 3.16). It is important to notice that this behavior was also found when cooling the films. A complete temperature cycle (up and down) was performed for a film of initial thickness $d_0 = 360$ Å, proving the reproducibility of the results.

The temperature range for the measurements of the freestanding PS films was limited to the region $25°\text{C} < T \leq 90°\text{C}$. For higher values, close to the bulk glass transition temperature $T_\text{G} \approx 100°\text{C}$, the fits of the reflectivity data reveal that the roughness increase drastically (from $\sigma_\text{PS} \approx 10$ Å to $\sigma_\text{PS} > 30$ Å), rendering an accurate determination of the film thickness virtually impossible. Investigations with an optical microscope show that holes begin to form, and finally, the films break down at higher temperatures.

For comparison measurements of the same kind were done with the control series of PS films supported by Si substrates. The relative thickness changes $d(T)/d_0$ obtained are depicted in Fig. 3.17. Again the thickest film ($d_0 = 863$ Å) shows bulk behavior. Below T_G the thermal expansion is weak because of the small thermal expansion coefficient, as mentioned above. For $T > T_\text{G}$ the expansion coefficient $\alpha_{\text{th.exp.}}$ jumps to a larger value and the increase of the film thickness with temperature is larger. A fit of the data (see the solid line in Fig. 3.17) in this region reveals a value of $\alpha_{\text{th.exp.}} \approx 5.4 \times 10^{-4}\,\text{K}^{-1}$, which is within the published range $\alpha_{\text{th.exp.}} \approx 5.1$–$6.0 \times 10^{-4}\,\text{K}^{-1}$ for bulk PS [46].

For thinner films, with thicknesses on the order of R_G, the same behavior as for the freestanding films is observed. The maximum contraction for the sample with $d_0 = 203$ Å was 3% (see Fig. 3.17). For the freestanding film with $d_0 = 360$ Å almost the same amount of contraction was found. Thus, the film contraction seems to be independent of the attractive interaction with a substrate. A significant difference between the freestanding film and the films on Si is that the substrate stabilizes the films and obviously prevents the formation of holes at higher temperatures. Therefore the PS films on Si could be measured over a larger temperature range, far above T_G.

Orts et al. [270] repot similar measurements on PS films spin cast on Si substrates. They explain the observed contraction by dewetting while the films were heated up [176, 177, 295]. The dewetting was confirmed by a significant increase of the surface roughness. Since dewetting can be ruled out in the case of the results shown in Fig. 3.17, because the fits of the reflectivity data reveal no significant increase of the surface roughness as a function of temperature [20], a different mechanism must be responsible for the contraction. The film contraction is most pronounced for the thinnest film with $d_0 = 45\,\text{Å}$. Figure 3.17 shows that this film contracts by about 15% (!!). Above $T_{G_{\text{bulk}}}$ only a very small expansion is observed. It should be noted that – similarly to the freestanding samples – the film regains its initial thickness after being cooled down to room temperature. This proves that equilibrium conditions were maintained during the experiments; this was further confirmed by measuring more temperature cycles, which yield identical results [19, 20].

This example shows how x-ray reflectivity gives insight into a phenomenon which could not have been measured otherwise. The physics behind this contraction effect in ultrathin PS films[22] is not clear yet, and more experiments with freestanding polymer films have to be performed to answer this question.

Also, the consequences of the film contraction on the glass transition temperature are not known. It is not at all clear whether this effect is connected with the glass transition or if it happens independently. One may speculate that the former is true and that mechanisms that yield an increase or decrease of T_G as discussed above [26, 234, 238] play a major role here, too. Hence the x-ray reflectivity experiments on the freestanding films indicate that the answer to the question of whether and how the glass transition is affected by the constrained geometry of a thin film is more complicated than simply "T_G decreases" or "T_G increases with decreasing film thickness".

3.2.2 Liquid Films on Solid Substrates

Liquid thin films may be considered as "liquids in confined geometry". How this "confined geometry" alters the morphology of the surface and the wetting behavior is of great practical interest for many applications, where of course painting is the most prominent. However, many of the more complicated wetting processes that occur in daily life are at best partially understood. Hence, a great amount of fundamental work on liquid thin films adsorbed on solid substrates has been done in the past.

[22] This contraction was also observed for a series of thin PEP films.

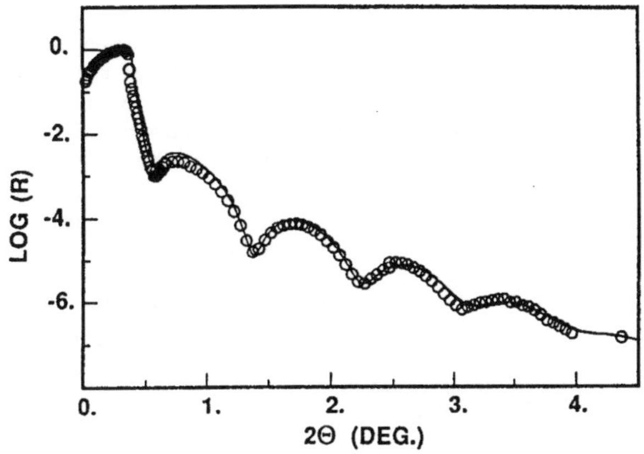

Fig. 3.18. Reflectivity (*symbols*) of a water film on a glass substrate in contact with a macroscopic meniscus (λ = 1.417 Å). The fit (*line*) with the density profile shown in Fig. 3.19 reveals a thickness of d = 91 Å, and roughnesses $\sigma_g \approx 7$ Å, and $\sigma_w \approx$ 4 Å (figure taken from *Garoff et al.* [135])

There are three major structural questions concerning the statics of films that completely wet a solid surface: (i) The influence of a solid substrate is expected to modify the solid/liquid interface. While this situation has been extensively studied theoretically in the past (see e.g. Refs. [3, 21, 56, 58, 86, 88, 107, 144, 151, 162, 229, 286, 380, 389, 403]), only a few experiments have been carried out so far. (ii) The liquid/vapor interface is also of fundamental interest. Calculations of the intrinsic liquid/vapor density profile have been done for many fluids (see e.g. Refs. [53, 58, 109, 110, 113, 169, 183]). Many aspects of capillary waves at the liquid/vapor interface and the question of how these capillary waves on liquid thin films are altered by the background potential due to the substrate have been the focus of a large body of theoretical work, too (see e.g. Refs. [29, 30, 53, 92, 93, 95, 130, 131, 241, 256, 259]). (iii) Further, one may ask whether the contour of the underlying solid affects the roughness of the liquid film, i.e. the question of roughness replication is of great interest. Here again, mainly theoretical work has been done in the past [14, 208, 298] and only a few experiments have been carried out [98, 365]. These questions are the major topics of this section, with points (i) and (ii) in the center. The third issue cannot be investigated by x-ray reflectivity measurements and will be discussed later in Sect. 6.2.4.

It is quite difficult to investigate liquid thin films on an angstrom scale. On the one hand they are difficult to prepare, and on the other hand probes such as AFM and STM cannot be used to get structural information. The preparation of these films requires sample cells which maintain a constant temperature with an accuracy on the order of mK over at least one day in an environment essentially filled with the vapor of the film material [96, 98, 99, 250, 251, 365, 366]. Therefore only a few studies exist in the literature. Some of them will be discussed here.

The first reported x-ray study of liquid films was carried out in 1989 by *Garoff et al.* [135]. These authors investigated the structure of static precursor

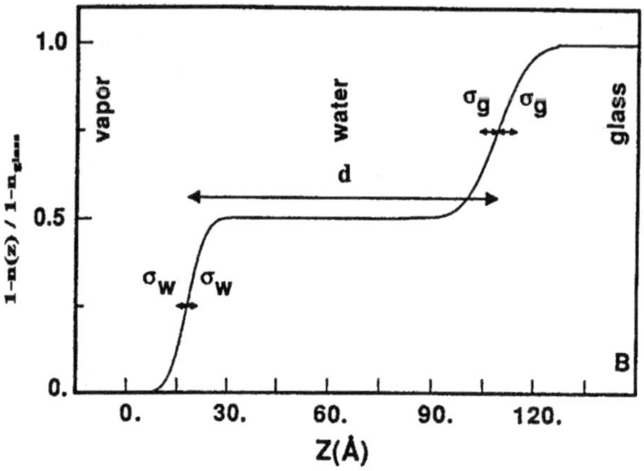

Fig. 3.19. Density profile corresponding to the reflectivity of the water film shown in Fig. 3.18. The water/glass and water/vapor interfaces are modeled by error-function profiles with interface widths (roughnesses) σ_g and σ_w (figure taken from *Garoff et al.* [135])

films preceding the macroscopic meniscus of water on glass. The symbols in Fig. 3.18 are the measured x-ray reflectivity of a film with thickness $d = 91$ Å. The interferences indicate that a homogeneous film was present during the examination. The fit to the data (line) was obtained with the density profile shown in Fig. 3.19. Two interface widths (roughnesses) were refined. Rms roughnesses of about $\sigma_g \approx 7$ Å for the glass/water interface and $\sigma_w \approx 4$ Å for the water/vapor interface were observed. This first examination showed the possibility of investigations of liquid thin films by x-ray reflectivity and demonstrated the sensitivity of the method.

In 1991 *Tidswell et al.* [365, 366] presented an x-ray study of thin cyclohexane films on silicon. The reflectivities are depicted in Fig. 3.20. The main emphasis was to show that very thin films with thicknesses $d < 60$ Å follow the roughness of the substrate conformally whereas the conformality is lost for thicker films of $d \sim 100$ Å. Neither the interface profiles at the substrate could be obtained from the data nor a systematic study of the vapor/liquid interface was performed. Since off-specular scattering was mainly the topic, this example is re-discussed later (see Sect. 6.2.2).

After these pioneering works, other liquid thin film systems were investigated with the x-ray reflectivity method. *Lurio et al.* [225, 226] determined the precise liquid/vapor interface profile of ultrathin He films. Their results will be discussed in Sect. 4.1.1 in terms of a kinematical approach. Later *Seeck et al.* [330] studied thin CCl_4 films, and *Strzelczyk et al.* [252, 360] investigated the growth of liquid CCl_4 and CCl_3Br films on various substrates. However, the data of these studies were somewhat inconclusive concerning a possible evolution of the surface roughness with the film thickness. The reason for this may be that the film thickness was controlled via a small temperature difference ΔT between the vapor and the liquid on the wafer, and thus the conditions were slightly away from thermal equilibrium.

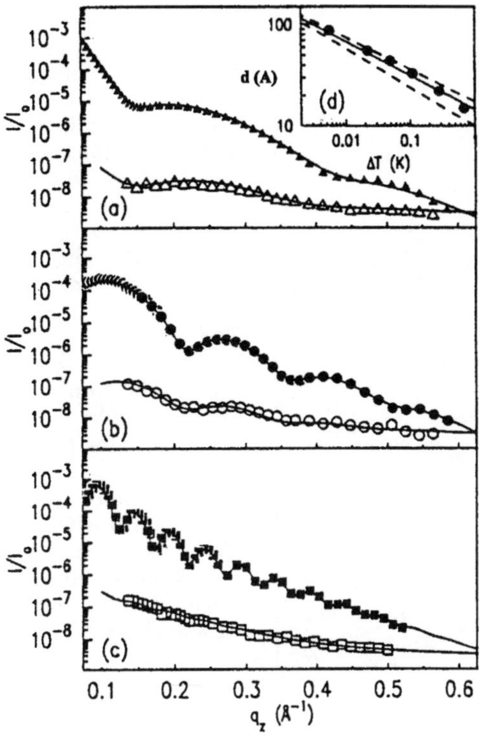

Fig. 3.20. Specular (*solid symbols*) and off-specular ($q_x/q_z = 0.0026$, *open symbols*) scattering of thin cyclohexane films on silicon substrates. The corresponding film thicknesses are **(a)** $d = 22$ Å, **(b)** $d = 44$ Å, and **(c)** $d = 128$ Å. The *lines* are fits to the data. The oscillations in the off-specular scans in (a) and (b) indicate roughness replication from the substrate, while no off-specular oscillations are visible for the thickest film (c). The inset **(d)** depicts a log–log plot of d vs. ΔT, the small temperature difference between the silicon wafer and the cyclohexane vapor which was introduced to control the film thickness. The *solid line* is a fit of the data (*solid circles*) while the *dashed lines* indicate the possible theoretical range (figure taken from *Tidswell et al.* [365])

The first experiments that yielded quantitative information about the two major open questions mentioned in the introduction of this subsection were performed by *Doerr et al.* [96, 98, 99]. The film thickness d was controlled via an isothermal pressure variation according to $p = p_{\text{sat}} \exp(-c/d^3)$, where c is a constant and p_{sat} is the pressure of the saturated vapor (Frenkel–Halsey–Hell equation [129]).

A logarithmic increase of the surface roughness with the film thickness, as typical for capillary waves (see Eq. 3.21), was observed and the precise solid/liquid interface profile was determined. As an example Fig. 3.21 shows the reflectivities of wetting hexane films with thicknesses in the range $d = 30$–350 Å (HASYLAB W1 beamline, $\lambda = 1.414$ Å). The decreasing modulation period indicates an increasing film thickness. A sophisticated sample cell provides the conditions for very stable films in the range $d = 30$–200 Å [99]. Only the thickest film of 350 Å was found to be unstable. From the experimental point of view it is very difficult to prepare thicker liquid films because they are extremely sensitive to temperature variations[23].

[23] A crude estimation shows that the maximum film thickness $d_{\max}$ is connected with the temperature stability δT via $d_{\max} \sim (\delta T)^{-1/3} \sim 300$–400 Å for realistic parameters [250, 365]. Also gravitation, limits $d_{\max}$ [135].

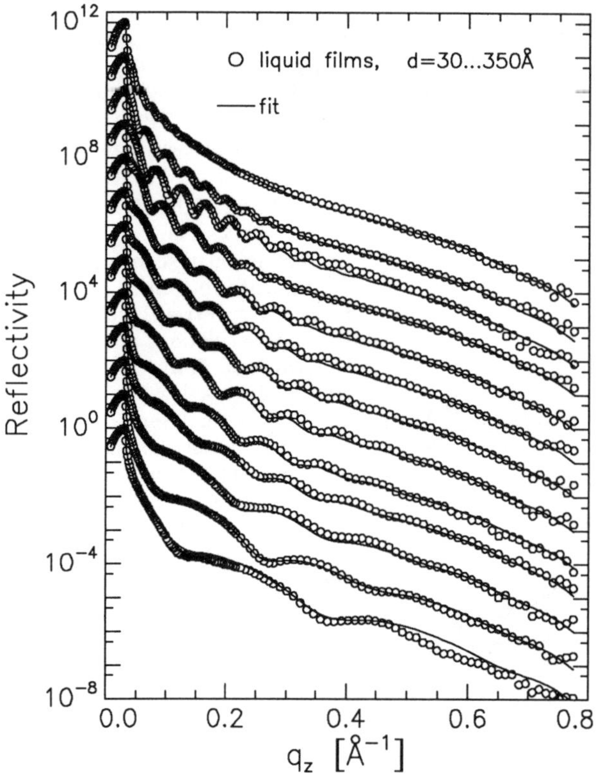

Fig. 3.21. Reflectivities (*symbols*) of a series of wetting hexane films on an oxide-covered silicon substrate. The film thicknesses were d = 30–350 Å (from *bottom* to *top*). The *lines* are fits to the data using the density profiles depicted in Fig. 3.22. The refinement was performed using the same solid/liquid interface profile for all films while for the liquid/vapor interface an error-function profile was assumed with variable width (roughness). The thickest film was excluded from further data analysis [96]

First, a simple slab model assuming a homogeneous hexane layer with electron density $\varrho_{\rm hex} = 0.236\,e/\text{Å}^3$, a thin oxide layer with $\varrho_{\rm ox} = 0.66\,e/\text{Å}^3$, and the silicon substrate with $\varrho_{\rm Si} = 0.697\,e/\text{Å}^3$ (the bulk values in electrons per cubic angstrom), was refined to explain the reflectivity data. This simple model leads to poor results. It turns out that a more complex profile, which is shown in Fig. 3.22 is able to explain the data quantitatively. The way this complicated profile was obtained and a discussion of the uniqueness of this solution will be given in Sects. 4.1.2 and 4.2.

The inset in Fig. 3.22 shows a magnification of the solid/liquid interfacial region. This interface was assumed to be the same for all films (solid circles). There is no reason to believe that the solid/liquid interface profile should vary with the film thickness.

Plischke & Henderson [286] have calculated density profiles of fluids near a hard wall by solving numerically the inhomogeneous *Percus–Yevick* equation, together with an equation of *Lovett et al.* [223]. Recently, *Patra & Ghosh* [58] presented a density functional approach to obtain density profiles and radial pair correlation functions for this scenario. The two studies show theoretically

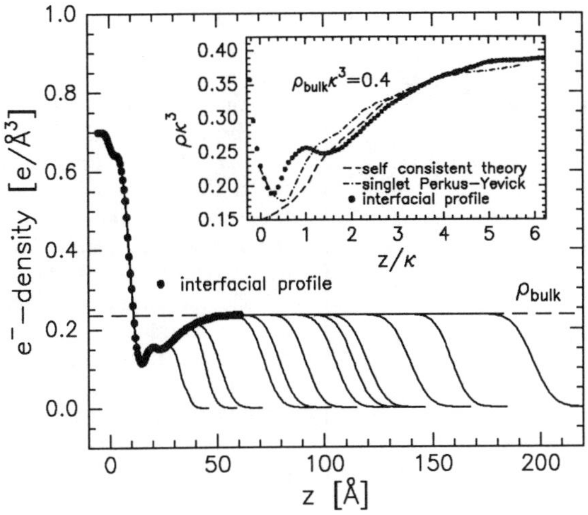

Fig. 3.22. Density profiles (*lines*) of thin liquid hexane films on an oxide-covered silicon substrate. The film thicknesses were $d = 30, \ldots, 200$ Å corresponding to the fits in Fig. 3.21. The solid/liquid interface (*symbols*) was assumed to be the same for all films (see magnification in the inset, and density functional calculations in Ref. [286]) [96]

that a region with considerably lower density than that of the bulk liquid may be induced by the constraint of the substrate.

In the study of *Plischke & Henderson* the density profile is obtained as a function of the dimensionless parameters $\hat{\varrho} = \varrho\,\kappa^3$ and $\hat{T} = k_{\mathrm{B}}\,T/\varepsilon$, where ϱ and κ are the bulk density and the molecular size, respectively, and ε is the minimum energy of the assumed 6–12 Lennard–Jones interaction potential. In general, the density profiles which were calculated vary from non-oscillatory profiles with a contact density far below that of the bulk, to profiles displaying oscillations and a contact density far above the bulk value. The dashed and dashed–dotted lines in the inset of Fig. 3.22 are the profiles according to the results of *Plischke & Henderson* [286] with $\hat{\varrho} = 0.4$ and $\hat{T} = 1.35$. Good quantitative agreement between the measured and the calculated profile is obtained for $\kappa \simeq 11$Å, a value relatively close to the length $\kappa = 9$ Å of a hexane molecule.

The low-density region near the substrate suggests that the packing density of the molecules is considerably reduced by the influence of the wall. Calculations by *Vossen & Forstmann* [389] show that polar molecules such as water tend to order vertically near an impermeable wall. Hence, a one-dimensional ice-like interface structure with reduced density is obtained. This situation cannot be transferred to hexane since the molecule is nonpolar. However, one may speculate that vertical order is indeed induced by the substrate because of the complex shape of a hexane molecule, which allows quite different packing densities. Investigations of hexane monolayers on graphite by *Krim et al.* [196] and *Newton* [262] have confirmed this picture: unexpected one-dimensional melting occurs, which is explained by the anisotropic complicated shape of the molecule.

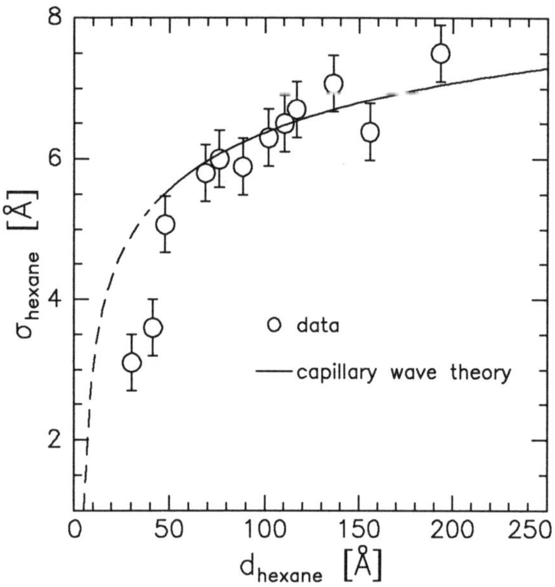

Fig. 3.23. Rms roughness σ_{hexane} vs. film thickness d_{hexane} corresponding to the measurements of the liquid films shown in Fig. 3.21. The *line* is a fit with a simple capillary wave model including a van der Waals cut-off $q_{\text{vdW}}(d)$, which leads to a logarithmic increase of the surface roughness with the film thickness (see Eqs. 3.20 and 3.21). The *dashed line* for smaller thicknesses depicts the region where this model is not applicable (see density profile in Fig. 3.22) [96]

Recent investigations of thin liquid decane and perfluoro-hexane films on silicon wafers reveal qualitatively the same results: a (less pronounced) low-density region is obtained close to the substrate. On the other hand, for spherical nonpolar molecules such as cyclohexane and CCl_4 on silicon, an enhanced density in the near-substrate region is found [99, 276] in accordance with various theoretical predictions (see e.g. Ref. [144]).

Now the results for the liquid/vapor interface of the hexane system will be further discussed. The density tail of the liquid/vapor interface has been modeled by an error-function profile. In principle, this profile represents both the intrinsic liquid/vapor profile and the profile due to thermally excited capillary waves. The latter are expected to dominate for bulk liquids and liquid films. This is confirmed by Fig. 3.23. The width σ_{hexane} of the interface as a function of the film thickness d_{hexane} is depicted. Since the films are not as thick as in the case of polymer films, Eq. (3.21) from the previous subsection is applicable[24]. For $d_{\text{hexane}} \geq 50$Å the logarithmic increase of the roughness as predicted by Eq. (3.21) (solid line) agrees well with the data (open circles). For thinner films (dashed line) this should not be the case, since the density profile in Fig. 3.22 reveals that pure capillary waves cannot be expected. Calculations by *Mecke & Dietrich* [241] predict for this thickness region a considerably lowered surface roughness, as indicated by the data in Fig. 3.23. The obtained values for the parameters B and a (see Eqs. 3.20 and 3.21) are $B = (14 \pm 2)$ Å^2 and $a = (10 \pm 2)$ Å. These values are equivalent to a surface

[24] The lower cutoff is always equal to the van der Waals cutoff $q_{\text{vdW}}(d) = a/d^2$, hence the roughness is resolution and q_z independent.

tension lowered by almost a factor of two compared to the bulk value[25]. The lower surface tension for thin films compared to the respective bulk value was also noticed for the PS and P2VP films in the previous section, where a modified interaction due to the entanglement of the long-chain molecules was able to explain the finding. The situation is now more subtle since this effect can be excluded for the short hexane molecules [96]. We will come back to these films in Chap. 6 and Sect. 6.2.2. Therefore the discussion is stopped here. It will be shown to what extent information about their lateral strucutre can be extracted from the diffusely scattered intensity.

3.2.3 Liquid Films on Liquid Surfaces

Another class of soft-matter thin films consists of those on bulk liquid substrates. An extensive review of diffraction experiments from organic molecules on water surfaces can be found in *Als-Nielsen et al.* [12]. Organic layers of polar molecules with amphiphilic head groups on water surfaces show a huge variety of phases depending on the chain lengths, head groups, and thermodynamic parameters. In the past x-ray scattering was an almost unique tool for structural investigations of these systems. Also, an investigation of monolayer surfactants on liquids down to nanometer scales was recently published by *Gourier et al.* [127, 146] yielding very interesting results concerning bending effects in these films.

Magnusson et al. [230] and *Regan et al.* [281, 293] have shown that even at free bulk liquid surfaces of mercury and gallium a layering may be observed. They performed x-ray reflectivity measurements over a wide q_z range, and found a peak which corresponds to a near-surface "layer" of ordered molecules in the liquid phase, which slightly changes the electron density profile. The thickness of this "layer" depends on the difference between the temperature of the liquid and the respective melting point. For metal surfaces this phenomenon was predicted theoretically long time ago, but experimentally it was not accessible until synchrotron sources provided hard x-ray beams of sufficiently high flux. Qualitatively, the layering at metal surfaces is induced by the abrupt change of the potential at the liquid/vapor boundary, quite comparable to the abrupt change introduced by a solid substrate as discussed in the preceding subsection. Further x-ray studies of gallium surfaces were focused on an understanding of the oxidation process on a microscopic level [285, 294].

Two examples of liquid soft-matter films on liquid substrates will be discussed in more detail: (i) diblock copolymer monolayers on water surfaces, and (ii) the surface crystallization of liquid normal alkanes ("surface freezing"). These examples were chosen because they show rather unexpected phase transitions[26].

[25] The upper cutoff was set to $q_{\mathrm{u,c}} = 2\pi/\kappa \approx 0.6\,\text{Å}^{-1}$.

[26] It should be noted that for these two examples off-specular x-ray measurements were also performed, which will not be discussed here [209, 400].

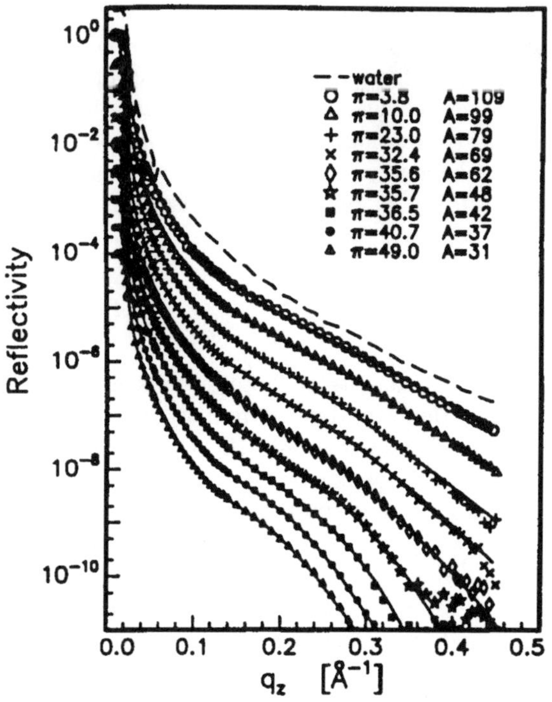

Fig. 3.24. Reflectivities (*symbols*) and fits (*lines*) with a two layer model for a PS-P4VP-C_nI^- monolayer film with $n = 8$ on a water surface. A two-layer model was chosen because it is assumed that a homogeneous PVP layer with PS micelles on top is present on the water surface (see Fig. 3.25). The *dashed line* shows for comparison the reflectivity of a water surface without polymer layer. The surface pressure Π increases from the *top* to the *bottom*, hence decreasing the area per molecule A. The curves are shifted on the intensity scale for clarity [209]

Recently diblock copolymers have been found to self-assemble at the air/water interface. They form circular structures referred to as surface micelles, with length scales on the order of 5–100 nm (see e.g. the works of *Israelachvili* [171] or *Zhu et al.* [416, 417]).

Figure 3.24 shows the reflectivities of a diblock copolymer monolayer (here PS-P4VP-C_nI^- with $n = 8$) on a water surface for various surface pressures Π corresponding to different areas per molecule A [209]. The Π–A isotherm reveals a plateau region for $50\,\text{nm}^2 < A < 70\,\text{nm}^2$, suggesting that a phase transition takes place. This is confirmed by the analysis of the specular-reflectivity data (see lines in Fig. 3.24). The fits are based on the density profile presented in Fig. 3.25. The thickness of the polymer layer increases only by a factor of two as the polymer surface density increases fourfold while being compressed. No significant hydration of the layer is observed. A detailed study for different PS-P4VP-C_nI^- polymers with $n = 1, 4, 8, 10$ was given by *Li et al.* [209]. However, since the x-ray reflectivity only corresponds to a density profile no information about the lateral structure can be obtained from the density profile shown in Fig. 3.25. Hence, diffuse scattering is needed to complete the picture. A peak was found corresponding to a lateral distance between 150 Å and 400 Å depending on the surface pressure. This proves the structure of the laterally ordered phase as sketched in the upper part

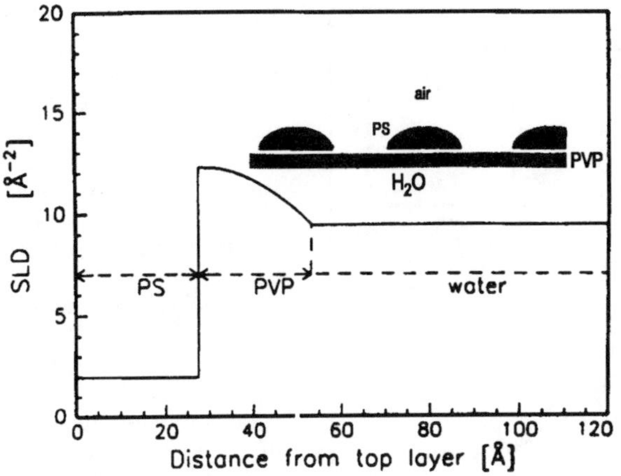

Fig. 3.25. Electron density profile (SLD: scattering length density $r_e\varrho(z)$) of a self-assembled diblock copolymer monolayer of PS-P4VP-C_8I^- on a water surface. The sketch shows the lateral arrangement of the polymer consistent with the density profile and with additional off-specular scattering data [209]

of Fig. 3.25. It is interesting to note that this phase can be transferred to silicon wafers without destroying the lateral order [210]. These self-assembled lattice structures consisting of diblock copolymer nano patterns may have industrial applications in the future since patterning on length scales of only a few hundred angstroms is achieved – a length scale not accessible with conventional methods of optical lithography.

The second example, presented in more detail, deals with liquid alkanes. It is well known that normal alkanes, i.e. linear hydrocarbon chains CH_3–$(CH_2)_{n-2}$–CH_3, also denoted as Cn, exhibit a rich variety of phases depending on the temperature and the chain length [341]. *Wu et al.* [400] presented a surface x-ray scattering study of n-alkanes in 1993. Figure 3.26 depicts the reflectivities of C18, C20, and C24. About 3°C above the bulk solidification temperature T_m a surface layer with larger density is formed [400, 401, 402]. This dense surface layer was identified as a new surface phase that occurs prior to solidification in a certain temperature range. The reflectivities presented in Fig. 3.26 also show that an increase of the surface roughness was not detected. GID experiments proved that the molecules of this dense surface layer correspond to a hexagonally close packed long-range-ordered structure where the molecules are oriented normal to the surface [400, 401]. A further investigation of this phenomenon shows that a single solid monolayer persists down to T_m. This surface ordered phase is obtained only for chain lengths $14 < n \leq 50$. It is quite insensitive to impurities and also was found for alkane mixtures and alcohols [97]. The vanishing of the phenomenon below $n = 14$ is interpreted as a possible transition from "surface freezing" to surface melting behavior[27]. In contrast to many other surface phase transition phenomena, surface freezing of n-alkanes was not theoretically predicted. Thus, this is an

[27] A prominent example of surface melting is ice, where *Lied, Dosch & Bilgram* [216] found that the surface layers already start to melt at $T_s = -18$°C.

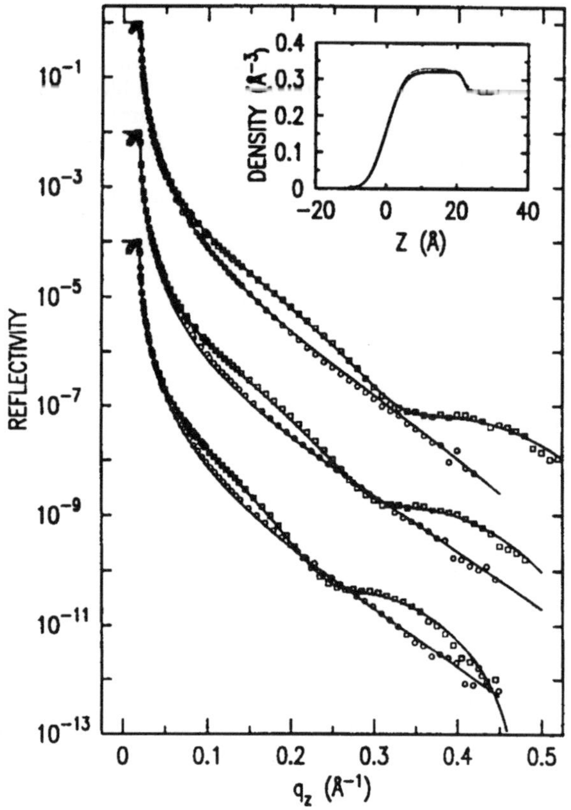

Fig. 3.26. Reflectivities of the normal alkanes C18, C20, and C24. The data for the liquid phase are shown by *open circles* and those for the surface monolayer phase by *open squares* where the oscillation is visible. The *lines* are fits to the data with the density profile displayed in the inset. The depletion layer is clearly seen. A ratio of the densities of the bulk material and the surface layer $\varrho_{\text{bulk}}/\varrho_{\text{surf.}} = 0.83$ is found, with a surface roughness of $\sigma \approx 4.0$ Å (liquid phase $\sigma \approx 4.5$ Å). Additional off-specular scans reveal that the surface monolayer is hexagonally close packed and long-range ordered (figure taken from *Wu et al.* [400])

example of a fundamental physical property discovered by x-ray reflectivity measurements.

Whereas the examples discussed up to this point have focused on the properties of single layers of soft-matter thin films, in the next subsection organic multilayers are considered.

3.2.4 Langmuir–Blodgett Films

Organic multilayers nowadays play an important role in thin-film technology research. Possible applications for these films are coatings on glass fibers for light transmission, high-speed optical control elements in microelectronics, improvement of the surface quality of mirrors, and detectors for organic molecules as biosensor devices [5, 132, 353].

Organic multilayers prepared by transferring molecules from a liquid subphase to a solid substrate by repeating a simple dipping procedure are called Langmuir–Blodgett (LB) films [5, 42]. This technique leads one to expect that imperfections will be transferred completely from one layer to the next,

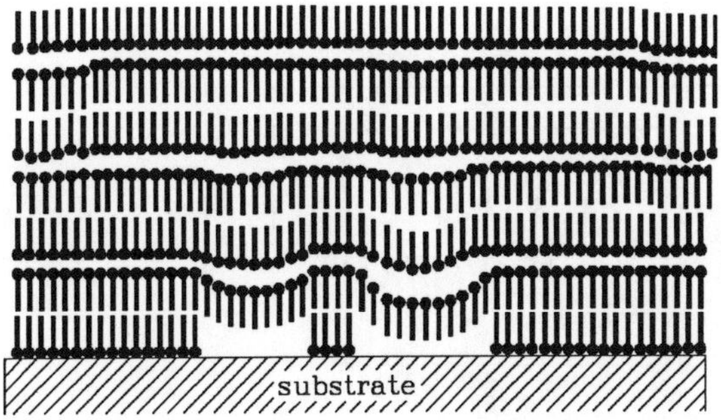

Fig. 3.27. Sketch of a Langmuir–Blodgett multilayer Y-structure (see inset of Fig. 3.29). *Solid circles*: polar head groups. *Straight lines*: organic tails. In the *middle*, vertical and lateral disorder is depicted. The disorder propagates to the layers above (conformal roughness)

yielding strong conformal or correlated roughness, i.e. disorder, in these structures. Such disorder is of decisive importance for future sensors based on LB films, since high-quality organic multilayers are required for all applications. Hence, techniques are needed to detect structural imperfections so that afterwards the preparation process may be optimized.

The vertical structure of an LB film is shown schematically in Fig. 3.27. The polar head groups of the molecules are drawn as bold circles and the organic hydrophobic tails as straight lines. At the left and right an almost perfect layer stack is present, whereas in the middle a defect is sketched. The defect propagates from the bottom to the top layer, causing conformal roughness, while keeping the thickness of the double layer almost constant. It was detected by GID measurements that the molecules are laterally ordered in domains of size 100–1000 Å [263, 264, 356, 357, 358].

Nitz et al. [264] confirmed this picture for cadmium arachidate (CdA) LB films. Figure 3.28 shows the reflectivity of an 11-layer CdA film together with an off-specular longitudinal diffuse scan with offset $\delta\alpha_{\mathrm{i}} = 0.05°$. The two curves are shifted for clarity by one order of magnitude on the intensity scale. The Bragg peaks of the one-dimensional periodic structure are visible. Their distance Δq_z in reciprocal space corresponds to a Cd–Cd spacing of $2\pi/\Delta q_z = 55\,\text{Å}$ as can be seen in the inset of Fig. 3.28. This confirms the so-called Y-structure, with the head groups on the surface of the substrate and alternate CdA bilayers (see inset of Fig. 3.29).

However, the reflectivity cannot be analyzed directly with the methods presented in Chap. 2. Transverse scans reveal that over a wide range, here $q_z > 0.5\,\text{Å}^{-1}$, the scattering is purely diffuse. One strategy would be to subtract the diffuse scattering and fit the remaining "true specular" data. This treatment is not accurate enough if such an amount of diffuse scattering is present. The fit in Fig. 3.28 (solid lines), with the density profile shown in the inset, was obtained in the opposite manner. Instead of subtracting the diffuse intensity, it was calculated at the specular condition $\alpha_{\mathrm{i}} = \alpha_{\mathrm{f}}$ and added to the *Parratt* reflectivity. This procedure has been successfully applied to many systems in the past (see e.g. Refs. [227, 323, 324, 330, 355]). The huge

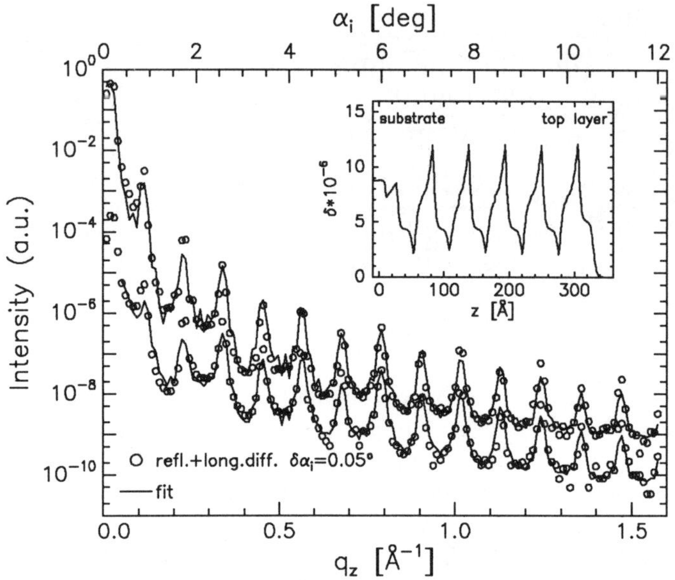

Fig. 3.28. Total reflectivity (specular plus diffuse) and a longitudinal diffuse scan with offset $\delta\alpha_i = 0.05°$ from an 11-layer cadmium arachidate LB film. The inset shows the corresponding $\delta(z)$ profile, i.e. the vertical structure, with the narrow Cd peaks [264]

amount of diffuse scattering, which replicates the features of the reflectivity almost perfectly, is a clear hint of strong vertical roughness correlations [244, 264, 314, 337, 338]. This will be further discussed in Sect. 6. In the study of *Nitz et al.* [264] no significant roughness increase from the bottom to the top inside the LB film was detected. Recently, *Lütt et al.* [228] presented a study of layer-by-layer grown film structures consisting of sequential depositions of oppositely charged macromolecules and "macrocycles" (ring-shaped molecules). Here an increase of the surface and interface roughness with the number of grown layers was detected and was explained with a simple random deposition model.

Results similar to those of *Nitz et al.* were found by *Gibaud et al.* [140], who investigated a sytem with nine CdA layers. Figure 3.29 depicts their reflectivity measurement, which again was recorded over a wide q_z range. Here the reflectivity in fact consists of diffusely scattered intensity for almost the entire q_z range, as confirmed by transverse scans. It should be noted again that none of the models presented in Chap. 2 are applicable in this case to fit the data! However, the Cd–Cd distance can be obtained from the spacing of the Bragg peaks as 55 Å, again confirming the Y-structure. Hence, these measurements also indicate the immense amount of roughness replication that is present in CdA films. The experiments of *Nitz et al.* and *Gibaud et al.* proved that the LB technique very efficiently transfers imperfections

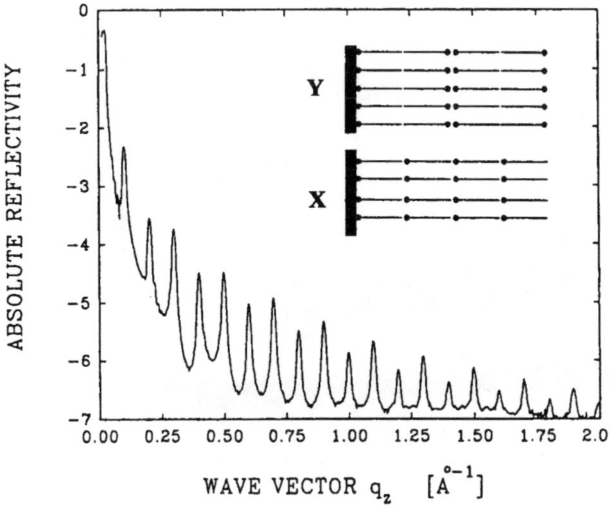

Fig. 3.29. Total reflectivity of a nine-layer cadmium arachidate LB film as obtained with a laboratory source. The regularly spaced Bragg peaks are arising from the multilayer structure. The X and Y configurations are shown in the inset. The reflected intensity is purely diffuse (figure from *Gibaud et al.* [140])

from one layer to the next. However, limited vertical roughness correlations of only three bilayers were found by *Barberka et al.* [24] in LB films of Cd stearate. Obviously the degree of roughness replication depends sensitively on the hydrophobic part of the molecules. This part is responsible for the bilayer–bilayer coupling which seems to be rather strong in CdA and less pronounced in Cd stearate LB multilayers.

The thermally induced rearrangement of molecules in Ba stearate LB films was recently investigated by *Englisch et al.* [111]. Figure 3.30 shows the x-ray reflectivities of a non annealed film (solid line, upper curve) and a film that was annealed to 65°C (dark squares, lower curve). Again the Bragg peaks were recorded over a wide range, indicating that the periodicity and average density of the Ba sheets are unchanged in the annealed film. Additionally, neutron reflectivity revealed that single molecules move across the multilayer and that this movement is enhanced by annealing, while the averaged structure is unchanged. A possible model for this scenario is sketched in Fig. 3.31. It depicts the structure of a Ba stearate multilayer, where the full circles are the polar bivalent head groups that connect two chains shown by the lines. The dotted lines indicate chains that have been deuterated. The data of *Englisch et al.* [111] suggest that with increasing temperature the van der Waals bonds to the neighboring chains break up. These unbridged chains are enabled to move almost unhindered across the multilayer, even on top of the original sample surface. However, no material is evaporated, since the critical angles of total-external-reflection for the x-ray (see inset of Fig. 3.30) and neutron reflectivity data were found to be almost unaffected by the annealing process.

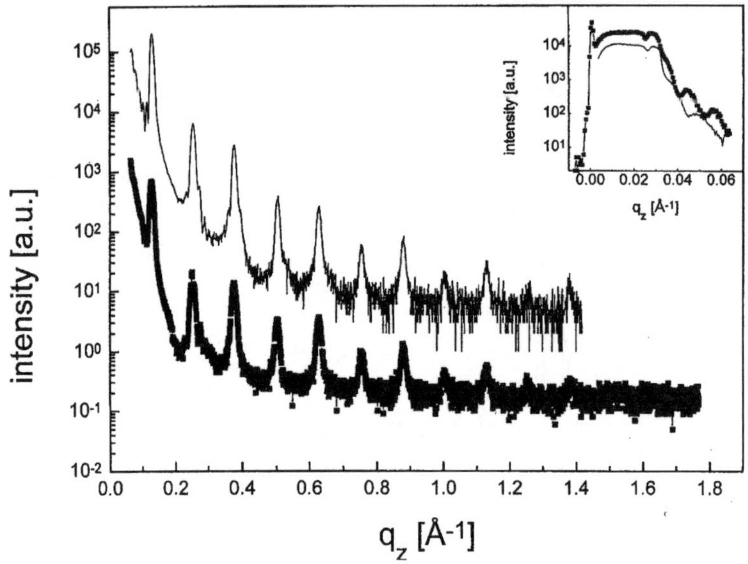

Fig. 3.30. X-ray reflectivity of a barium stearate LB film before annealing (*solid line, upper curve*) and after annealing (*squares, lower curve*). The inset shows the region close to the critical angle (from *Englisch et al.* [111])

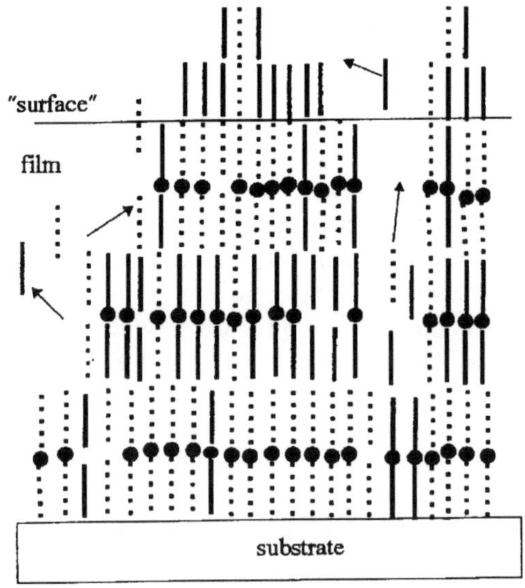

Fig. 3.31. Model of the vertical disorder of the molecules within a fatty-acid salt multilayer as obtained from x-ray and neutron reflectivity measurements. The *lines* are the chains (*dotted lines* indicate deuterated chains) and the *full circles* are the bivalent ions bridging two chains. The *arrows* indicate possible paths of molecular exchange (figure taken from *Englisch et al.* [111])

In the last decade x-ray and neutron reflectometry have become almost a standard tool to characterize LB films. However, the discussion of examples will be stopped here. The interested reader is referred to the numerous articles in the literature (see Refs. [9, 12, 167, 303, 304] and references therein).

At the end of this section, a few general remarks will be made: (i) As already mentioned, the investigated systems are examples where almost no method other than x-ray reflectometry can be applied to obtain detailed structural information on atomic up to mesoscopic length scales. Neither direct observations via modern microscopy techniques (AFM, STM, TEM) nor other scattering techniques such as ellipsometry would have been able to yield more reliable results from these systems. (ii) Another advantage of x-ray scattering is that the technique is non destructive. However, particularly for investigations of soft matter, this "standard argument" is not true in general. Radiation damage of samples may occur and data obtained at synchrotron sources have to be carefully checked with regard to reproducibility. (iii) A major disadvantage, however, of the x-ray reflectivity technique is the complicated data analysis: It is sometimes rather difficult, tedious, time consuming, and highly ambiguous.

The last point will be addressed in the next chapter from a more fundamental standpoint. The data analysis may be simplified by inversion techniques, and procedures are presented to minimize ambiguities (see Sects. 4.2 and 4.3).

4. Advanced Analysis Techniques

It was shown in the last chapter that x-ray reflectivity has been successfully applied to investigate many soft-matter materials. But one should always bear in mind that the density profiles obtained may not be unique, since they were obtained by fitting and not by a direct inversion of the data. In general, a reconstruction of a density profile from a single reflectivity measurement is impossible because phase information is lost by observing intensities rather than field amplitudes. This fact is well known as the "phase problem" of x-ray scattering.

However, quite often it turns out that this situation can be significantly improved. The electron density profile may indeed be fully determined by a single x-ray reflectivity measurement if certain conditions are fulfilled. Therefore, x-ray reflectivity is re-discussed in the next section from a more theoretical standpoint, where the simple kinematical approximation is in the center of the considerations. The main focus is to develop new, advanced analyzing techniques that avoid the time-consuming data fitting described in Sect. 3.1.2. In Sect. 4.2 an inversion technique will be presented that is mainly based on a "good initial guess" of the unknown phase of the reflection coefficient. Finally, in Sect. 4.3 other inversion techniques based on anomalous scattering are briefly outlined.

4.1 The Kinematical Approximation

In Chap. 2 reflectivity was discussed in terms of an optical language. The basic results that a single surface reflects according to the Fresnel formula (see Eq. 2.12) and that the reflectivity from arbitrary density profiles $\varrho(z)$ may be calculated by the slicing method via the *Parratt* formalism (see Sects. 2.2 and 2.4) are quite simple, and in fact independent of the assumption that hard x-rays with wavelengths on the order of $\lambda \sim 1$ Å are involved in the scattering process. The above-mentioned basic formulas would also apply in the case of soft x-rays or even visible light and can be found in optics textbooks such as the classic book of *Born & Wolf* [44]. Now we will focus on a different

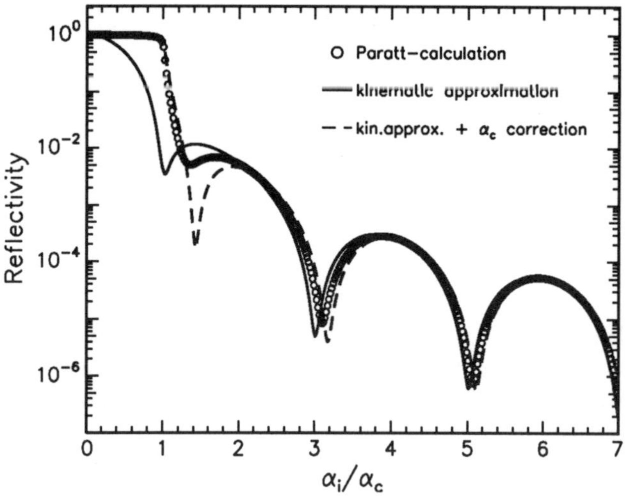

Fig. 4.1. Comparison between the exact *Parratt* reflectivity (*open circles*) and the kinematical result (*solid line*) for a polymer film of 100 Å thickness on a silicon substrate. The *dashed line* is a semi kinematical calculation taking refraction corrections into account

description that is more common in the x-ray scattering community: the kinematical approximation[1].

Although the formulas derived in Chap. 2 are quite useful for applications, the kinematical approximation allows a clearer treatment of the scattering, from which general conclusions may be more easily drawn (see Sects. 4.1.1 and 4.1.2). The kinematical approach is valid in the so-called "weak-scattering regime", i.e. when the cross section for the scattered radiation is small, and hence multiple-scattering effects may be neglected. This means that the kinematical approximation breaks down in the region of the critical angle, while it is valid for incident angles of, say, $\alpha_\mathrm{i} > 3\,\alpha_\mathrm{c}$, since the reflected intensity rapidly decreases for $\alpha_\mathrm{i} > \alpha_\mathrm{c}$ (see Eq. 2.12, and Figs. 2.2 and 4.1)[2].

A well known result is that if only single-scattering events need to be considered, the cross section is proportional to the Fourier transform of the total three-dimensional electron density $\varrho(x, y, z)$ of the scatterer. For surfaces this can be reformulated, and the following expression results [9, 11, 12, 85, 280]:

$$R(q_z) = R_\mathrm{F}(q_z) \left| \frac{1}{\varrho_\infty} \int \frac{\mathrm{d}\varrho(z)}{\mathrm{d}z} \exp(\mathrm{i}\, q_z z)\, \mathrm{d}z \right|^2 = R_\mathrm{F}(q_z) \left| F(q_z) \right|^2 \qquad (4.1)$$

for the reflectivity $R(q_z)$. Here $q_z = 2k \sin \alpha_\mathrm{i}$ denotes the vertical wavevector transfer, $\varrho(z)$ is the laterally averaged electron density profile of the sample[3], and ϱ_∞ is the average density of the entire sample. The pre-factor $R_\mathrm{F}(q_z)$ is the Fresnel reflectivity of the substrate given by Eq. (2.12), and the structure

[1] Also known as the (first-order) Born approximation.

[2] Multiple-scattering effects also dominate in the center of Bragg reflections.

[3] It should be noted again that this corresponds directly to a dispersion or refractive-index profile, $\delta(z)$ or $n(z)$, respectively, as discussed in Chap. 2.

factor $F(q_z)$ is the Fourier transform of the derivative of the electron density profile $\varrho(z)$.

It is simple to show that Eq. (4.1) is, in a sense, equivalent to the exact optical formalism developed in Chap. 2, if the incident angle is much larger than the critical angle. This can also be seen in Fig. 4.1 where the reflectivity from a polymer film of thickness 100 Å on a silicon substrate ($\delta_{\mathrm{Si}} = 7.56 \times 10^{-6}, \delta_{\mathrm{PS}} = 3.5 \times 10^{-6}$ for $\lambda = 1.54$ Å) is shown (see Eq. 4.2 below). The symbols are the exact reflectivity calculated with Eq. (2.16) according to the *Parratt* formalism. The solid line is a kinematical calculation using Eqs. (4.1) and (4.2). For $\alpha_{\mathrm{i}} > 3\,\alpha_{\mathrm{c}}$, i.e. $R(q_z) < 10^{-3}$ ("weak scattering"), the two curves are almost identical. Large differences are present in the region of the critical angle, as $R \simeq 1$, there is "multiple scattering". The difference is slightly reduced by introducing a refraction correction for incident angles $\alpha_{\mathrm{i}} > \alpha_{\mathrm{c}}$. If q_z in the structure factor $F(q_z)$ of Eq. (4.1) is replaced by $q_z' = 2k\sin[(\alpha_{\mathrm{i}}^2 - \alpha_{\mathrm{c}}^2)^{1/2}]$ [11, 12] the dashed line in Fig. 4.1 results[4]. However, this approximation also fails in the vicinity of the critical angle but is almost identical to the exact result and the uncorrected approximation for $\alpha_{\mathrm{i}} > 3\,\alpha_{\mathrm{c}}$.

The reflectivity of a particular system can be obtained directly from Eq. (4.1). For a sharp interface located at $z = 0$ one finds $\mathrm{d}\varrho(z)/\mathrm{d}z = \delta(z)$, and thus $F(q_z) = 1$ and $R(q_z) = R_{\mathrm{F}}(q_z)$. If this interface is rough, with an error-function profile of width σ, one gets $\mathrm{d}\varrho(z)/\mathrm{d}z \sim \exp\{-z^2/(2\sigma^2)\}$, and thus $F(q_z) = \exp(-q_z^2\sigma^2/2)$ and $R(q_z) = R_{\mathrm{F}}(q_z)\exp(-q_z^2\sigma^2)$, i.e. the *Beckmann–Spizzichino* result given by Eq. (2.40) [31].

If we consider the case of a single layer of thickness d on top of a substrate with independent profiles $\varrho_{1,2}(z)$ at the interfaces, the structure factors $F_{1,2}(q_z) = (1/\varrho_\infty)\int \mathrm{d}\varrho_{1,2}(z)/\mathrm{d}z \, \exp(\mathrm{i}\,q_z z)\,\mathrm{d}z = |F_{1,2}(q_z)|\exp\{\mathrm{i}\,\phi_{1,2}(q_z)\}$ with the phases $\phi_{1,2}(q_z)$ may be introduced, and Eq. (4.1) yields

$$R(q_z)/R_{\mathrm{F}}(q_z) = \left|F_1(q_z) + \exp(\mathrm{i}\,q_z d)\,F_2(q_z)\right|^2 = \left|F_1(q_z)\right|^2 + \left|F_2(q_z)\right|^2 + 2\left|F_1(q_z)\right|\left|F_2(q_z)\right| \cos\left\{q_z d + \Delta\phi(q_z)\right\}, \tag{4.2}$$

with $\Delta\phi(q_z) = \phi_1(q_z) - \phi_2(q_z)$. The last term in Eq. (4.2) is responsible for the interference between x-rays reflected from the top and bottom of the layer (so-called "Kiessig fringes" [188]). More complex systems may be treated in a similar manner. In Sect. 2.3 it was noticed that the asymmetry of a profile has no significant influence on the reflectivity of a single surface (see Eqs. 2.52–2.58 and Figs. 2.13 and 2.14). As an example showing that the kinematical approximation provides more insight into such fundamental questions about the reflectivity, asymmetric profiles are the topic of the next subsection.

[4] Here it must be noted that for thick layers the critical angle $\alpha_{\mathrm{c,film}}$ of the film rather than that of the substrate has to be used.

4.1.1 Asymmetric Profiles: A Closer Look

The reason why an asymmetry of a density profile in the case of a single surface is not detectable with x-ray reflectivity becomes clear from the considerations below. A very extensive discussion of scattering from non-Gaussian rough surfaces was recently published by *Zhao, Wang & Lu* [415]. Here we follow the description given by *Press et al.* [288][5].

For one interface a cumulant expansion of $F(q_z)$ in Eq. (4.1) yields

$$F(q_z) \approx \exp\left(\mathrm{i}q_z K^{(1)} - q_z^2 K^{(2)}/2 - \mathrm{i}q_z^3 K^{(3)}/6 + q_z^4 K^{(4)}/24 + \ldots\right), \quad (4.3)$$

where $K^{(m)}$ denotes the cumulant of order m. These cumulants are related to the moments $M^{(m)} = \int z^m P(z)\,\mathrm{d}z$ of a distribution $P(z)$ (here $P(z) = \mathrm{d}\varrho(z)/\mathrm{d}z$) in the following way:

$$M^{(1,2,3)} = K^{(1,2,3)}\,, \qquad M^{(4)} = K^{(4)} + 3K^{(2)2}\,. \quad (4.4)$$

The relation between moments and cumulants up to the tenth order is given by *Kendall* [186]. The first cumulant describes the mean position μ of the interface, which may be normalized to $\mu = 0$. The term of order $m = 2$ is equivalent to the usual rms roughness σ. In the expansion given by Eq. (4.3) the $m = 2$ term yields the *Beckmann–Spizzichino* damping factor stemming from an error-function interface profile (see Sect. 2.3). Higher order terms ($m \geq 3$) describe deviations from this profile, i.e. from a Gaussian height distribution. The term with $m = 3$ is related to an asymmetry of this height distribution $P(z)$, and the $m = 4$ contribution accounts for symmetric deviations from the Gaussian shape[6].

Calculating the reflectivity via $|F(q_z)|^2$ makes it obvious that only cumulants of even order can be determined. Hence, the asymmetry of an interface cannot be obtained from a simple reflectivity measurement, since all odd cumulants have imaginary arguments in the exponentials. This demonstrates how phase information is lost. Equation (4.3) also shows that for $q_z\sigma < 1$ the reflectivity of an interface is always similar to that of an error-function profile. Symmetric deviations may only be detected if either the roughness σ or the wavevector transfer q_z is sufficiently large. This is the explanation of why all the reflectivities from single surfaces in Sect. 2.3 that are depicted in Figs. 2.10 and 2.14 look quite similar.

The situation changes in the case of a layer on a substrate. Here additional interferences occur, which may serve as source of phase information. This is shown in the general expression for the scattering from a layer, which is given by Eq. (4.2). If both of the scattering amplitudes $F_1(q_z)$ and $F_2(q_z)$ are expanded according to Eq. (4.3) then the interference term becomes

---

[5] Similar calculations but without cumulants are given by *Rieutord et al.* [297].

[6] In general, the distribution is "flatter" than a Gaussian if $K^{(4)} < 0$ and "steeper" if $K^{(4)} > 0$. It is not possible to reconstruct the entire profile from a limited number of cumulants.

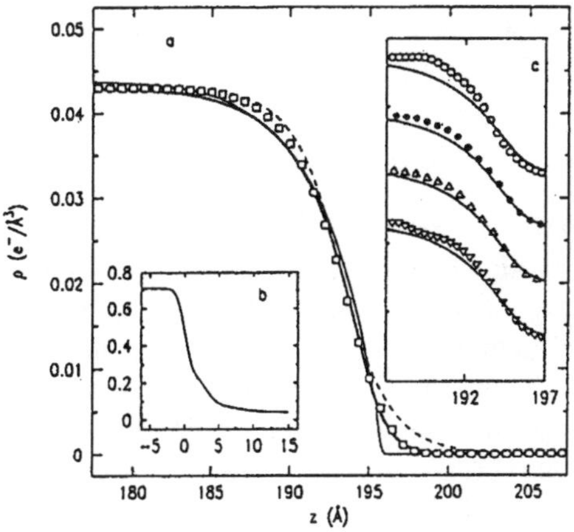

Fig. 4.2. (**a**) Density profile at the ^{4}He/vapor interface. The *squares* denote a theoretical profile [359], while the *solid line* in the middle is the profile from the best fit of the x-ray data. The other lines visualize the maximum and minimum asymmetry consistent with the data. (**b**) Profile at the ^{4}He/Si interface. (**c**) Comparison of other theoretical profiles with the best fit (from *Lurio et al.* [225])

$$\cos\left\{q_z d + \Delta\phi(q_z)\right\} = \cos\left\{q_z d - q_z^3(K_2^{(3)} - K_1^{(3)})/6\right\}, \tag{4.5}$$

where $K_1^{(3)}$ and $K_2^{(3)}$ are the $m = 3$ cumulants of both interfaces, and terms with $m > 3$ have been neglected. The dominant contribution is the modulation of the intensity with a period $\Delta q_z = 2\pi/d$, i.e. the *Kiessig* fringes. Using the difference $\Delta K^{(3)} = K_2^{(3)} - K_1^{(3)}$, one can redefine an effective periodicity $\Delta q_z = 2\pi/d(q_z)$, with $d(q_z) = d - q_z^2 \Delta K^{(3)}/6$. Evidently, the modulation period becomes q_z-dependent in the case of asymmetric profiles. However, one has to measure up to quite large wavevector transfers to observe a noticeable effect. A numerical example may be found in the paper of *Press et al.* [288].

As an example, the study of *Lurio et al.* [225] on thin liquid ^{4}He films will be discussed. Figure 4.2 depicts the density profile obtained from the x-ray reflectivity data. A convolution of a capillary wave term and an asymmetric tanh profile as introduced by Eqs. (2.52)–(2.58) was used for the refinement. The data are well explained with this assumption (see solid lines in Fig. 4.2). On one hand, this shows that x-ray reflectivity is a very sensitive tool to probe tiny features of density profiles when interference effects are used. On the other hand, Fig. 4.2 proves also that a clear decision between many possible density profiles is hard to achieve (see also the discussion by *Rieutord et al.* [297]) unless additional phase information is available. In Sect. 4.2 a way is outlined of how this additional information may be generated.

At present, no other experiment have been found where odd-order cumulants evidently contribute in the aforementioned way, via a q_z-dependent change of the modulation period of the Kiessig fringes. There are, however, some examples where the fourth order cumulant $K^{(4)}$ obviously plays a role. This is the case for polymer interdiffusion profiles in the near-surface regime where *Kunz & Stamm* [199, 200] introduced two Gaussians of different widths to explain reflectivity data. The underlying reptation model [100] for polymer diffusion requires at least two different diffusion constants. Here the introduction of a cumulant $K^{(4)}$ would represent an alternative.

4.1.2 Unique Profiles

Equation (4.1) describes the reflectivity of a given electron density profile $\varrho(z)$ in the kinematical approximation. It directly illustrates the major problem of x-ray scattering: The complete loss of phase information due to the fact that the structure factor $F(q_z)$ is a complex number and only the modulus squared of $F(q_z)$ enters the formula. Therefore the Fourier back transformation of Eq. (4.1) is impossible, and $\mathrm{d}\varrho(z)/\mathrm{d}z$ and $\varrho(z)$ cannot be directly and unambiguously obtained from a single reflectivity measurement. However, the situation is not as bad as it looks at first sight.

Considerations of the so-called Hilbert phase $\phi_{\mathrm{H}}(q_z)$ of analytic functions reveal that for certain cases the phase $\phi(q_z)$ of the structure factor $F(q_z)$ is identical to this Hilbert phase. This is important because $\phi_{\mathrm{H}}(q_z)$ is determined only by the modulus $|F(q_z)|$, i.e. by the reflectivity. In Appendix A.1 more details about the Hilbert phase are given. It turns out that $\phi_{\mathrm{H}}(q_z) \equiv \phi(q_z)$ if the density profile of an N-layer system with density contrasts $\Delta\varrho_m$ fulfills the following condition:

$$|\Delta\varrho_1| > \sum_{m=2}^{N} |\Delta\varrho_m| \,, \tag{4.6}$$

where $\Delta\varrho_1$ is the density contrast between the material of the semi-infinite substrate and that of the first layer [60]. Surprisingly, this condition is not very stringent and is often fulfilled, particularly for soft-matter thin films, where less dense materials lie on top of substrates with much larger electron densities.

However, even small roughnesses may change the situation totally. If a film of thickness d on a substrate is considered, where independent error-function profiles of widths σ_1 and σ_2 for the interfaces are assumed, then $F(q_z)$ is given by

$$F(q_z) = \Delta\varrho_1 \exp(-q_z^2\sigma_1^2/2) + \exp(\mathrm{i}\, q_z\, d)\, \Delta\varrho_2 \exp(-q_z^2\sigma_2^2/2) \,, \tag{4.7}$$

with $\Delta\varrho_1$ and $\Delta\varrho_2$ being the substrate/film and film/vacuum density contrasts, respectively. The particular example of a hexane film of thickness $d = 100\,\text{Å}$ on a silicon substrate will be discussed now. In this case $\Delta\varrho_1 = (0.699 - 0.235)\, e/\text{Å}^3 > \Delta\varrho_2 = 0.235\, e/\text{Å}^3$ and the criterion given by Eq. (4.6) is fulfilled. Therefore the exact phase $\phi(q_z)$ of $F(q_z)$ corresponding to the sharp profile and the Hilbert phase $\phi_{\mathrm{H}}(q_z)$ (see Appendix A.1, Eq. A.2) that can be calculated from $|F(q_z)|$, coincide.

But if roughnesses are introduced, the situation changes. For $\sigma_1 = 0$ (sharp substrate interface) and $\sigma_2 = 5\,\text{Å}$, Fig. 4.3 indicates that the two phases still agree. In the opposite case, $\sigma_1 = 5\,\text{Å}$ and $\sigma_2 = 0$, Fig. 4.4 reveals

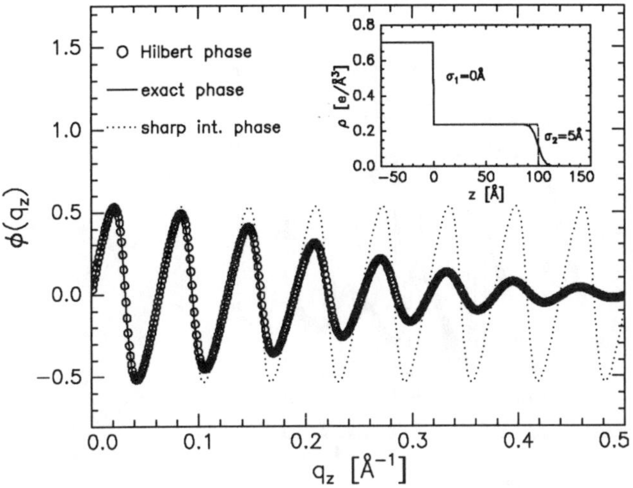

Fig. 4.3. Hilbert phase (*open circles*) calculated from the reflectivity only, and exact phase (*solid line*) for the electron density profile shown in the inset. The two phases are identical. The *dotted curve* is the phase for the *dotted profile* with sharp interfaces

that the phases disagree. It can be seen that the criterion given by Eq. (4.6) may not be valid in the case of rough interfaces.

This simple example shows why in practice an inversion of x-ray reflectivity data with the Hilbert phase $\phi_H(q_z)$ is of minor importance: Rather large ambiguities are still present. Even worse, if data are directly inverted using the Hilbert phase method, the obtained electron density profile is just one of maybe an infinite number of profiles that correspond to the data. Furthermore, the condition for $\phi(q_z) \equiv \phi_H(q_z)$ given by Eq. (4.6) does not help too much in reality: Since only a finite q_z range is covered by a measurement, additional ambiguities may result. The main disadvantage of a direct inversion using $\phi_H(q_z)$ is the neglect of all a-priori information about the system under consideration. In almost all cases this a-priori information is available, at least a rough estimate of the vertical structure of the sample is always known. How this a-priori information may be used to "guess" the unknown phase with the help of the Hilbert phase is the topic of the next section.

4.2 The "Phase-Guessing" Inversion Method

We start with the identity relating the derivative of the unknown density $\varrho(z)$ to the Fourier transform of the structure factor (see Eq. 4.1), i.e.

$$\varrho'(z) = \frac{\mathrm{d}\varrho(z)}{\mathrm{d}z} = \int F(q_z) \exp(\mathrm{i}\, q_z z)\, \mathrm{d}q_z \,, \tag{4.8}$$

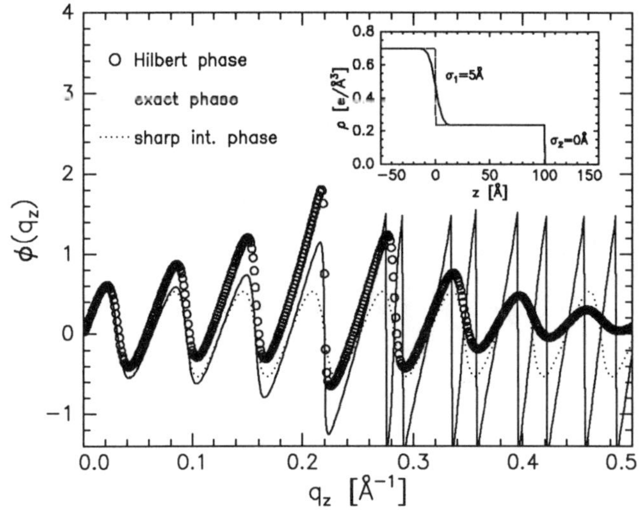

Fig. 4.4. Hilbert phase (*open circles*) determined by the reflectivity only, and exact phase (*solid line*) for the electron density profile for a rough substrate/adsorbate interface as shown in the inset. The two phases disagree. The *dotted curve* is the phase for the *dotted profile* possessing sharp interfaces

$$F(q_z) = \left|F(q_z)\right| \exp\left\{\mathrm{i}\,\phi(q_z)\right\} . \tag{4.9}$$

Constant pre-factors are neglected here. The central assumption is that crude estimates of the unknown system parameters are known so that an electron density profile $\varrho_m(z)$ can be "guessed". Of course this seems to be somehow a contradiction of the aim of a reflectivity measurement – the determination of the unknown density profile of a sample. However, in almost all cases the materials which are involved in the experiments are known beforehand, and hence their bulk densities are known. A trivial but important constraint is that all densities have to be positive numbers[7]. Moreover, the expected layer thicknesses may be known from other experiments or from a quick fit of the reflectivity data with the models outlined in Chap. 2. All these seemingly innocent pieces of pre-information can be used for a "guess" of the unknown phase and to reconstruct the density profile $\varrho(z)$ from a single reflectivity measurement. The method developed in this section was originally introduced using the DWBA by *Sanyal et al.* [22, 316], who later also applied the formalism within the more simple kinematical treatment [318, 319]. *Doerr et al.* [96] recently presented a new algorithm that includes a-priori information, as explained above, together with the Hilbert phase (see previous section and Eq. A.2).

[7] This does not hold for neutron reflectivity measurements, where negative scattering-length densities are possible.

It is assumed that such a "guessed" profile $\varrho_m(z)$ exists. Then similar equations to Eqs. (4.8) and (4.9) may be written down:

$$\varrho'_m(z) = \frac{\mathrm{d}\varrho_m(z)}{\mathrm{d}z} = \int F_m(q_z)\exp(\mathrm{i}\,q_z z)\,\mathrm{d}q_z\,, \tag{4.10}$$

$$F_m(q_z) = \left|F_m(q_z)\right| \exp\left\{\mathrm{i}\,\phi_m(q_z)\right\}, \tag{4.11}$$

where in general the (known!) phase $\phi_m(q_z)$ differs from the (unknown) phase $\phi(q_z)$, i.e. $\phi(q_z) = \phi_m(q_z) + \Delta\phi(q_z)$. Equations (4.8)–(4.11) together with Eq. (4.1) yield

$$\varrho'(z) = \iint \sqrt{\frac{R(q_z)}{R_m(q_z)}}\, \exp\left\{\mathrm{i}\,\Delta\phi(q_z)\right\} \varrho'_m(z_1) \exp\left[\mathrm{i}\,q_z(z - z_1)\right] \mathrm{d}z_1\mathrm{d}q_z\,, \tag{4.12}$$

which is still an exact expression for $\varrho'(z)$ and well-suited for a phase approximation.

If one assumes that the phase of the structure factor of the guessed profile and that of the system under consideration are essentially the same, then $\Delta\phi(q_z) \simeq 0$ and $\varrho'(z)$ may be calculated directly via Eq. (4.12). In practice, this crude approximation already leads to reasonable results [22, 318]. A better estimate for $\Delta\phi(q_z)$ is

$$\Delta\phi(q_z) \approx \phi_{\mathrm{H}}(q_z) - \phi_{\mathrm{H},m}(q_z)\,, \tag{4.13}$$

which is the difference between the known Hilbert phases. These phases correspond to the measured $R(q_z)$ and the calculated $R_m(q_z)$, respectively. If the phase difference given by Eq. (4.13) is used in Eq. (4.12) then a better convergence of the iterative density profile reconstruction algorithm given below is achieved [96]. The reason for taking $\phi_{\mathrm{H},m}(q_z)$ in Eq. (4.13), instead of the accessible exact model phase $\phi_m(q_z)$, is that ambiguities may cancel if both phases are calculated in the same manner.

Now the algorithm of the "phase-guessing" method can be described:

- (i) Measure the reflectivity $R(q_z)$ of the sample with an electron density profile $\varrho(z)$, and define an accuracy ε.
- (ii) "Guess" a profile $\varrho_m(z)$ from a-priori information or from a quick fit of $R(q_z)$.
- (iii) Generate $R_m(q_z)$ with the help of Eq. (4.1) or the *Parratt* formalism from this initial "guess".
- (iv) Generate the Hilbert phase difference $\Delta\phi(q_z) = \phi_{\mathrm{H}}(q_z) - \phi_{\mathrm{H},m}(q_z)$ via

$$\Delta\phi(q_z) = \frac{q_z}{\pi} \int\limits_0^\infty \frac{\ln\{[R_m(q'_z)R(q_z)]/[R_m(q_z)R(q'_z)]\}}{q'^2_z - q^2_z}\,\mathrm{d}q'_z\,.$$

- (v) Calculate

$$\varrho_{m+1}(z) = \int_{-\infty}^{z} \int_{-\infty}^{\infty} \int_{-\infty}^{\infty} \sqrt{\frac{R(q_z)}{R_m(q_z)}} \exp\left\{\mathrm{i}\,\Delta\phi(q_z)\right\} \times \frac{\mathrm{d}\varrho_m(z_1)}{\mathrm{d}z_1} \exp\left\{\mathrm{i}\,q_z(z_2 - z_1)\right\} \mathrm{d}q_z \mathrm{d}z_1 \mathrm{d}z_2 \,. \tag{4.14}$$

- (vi) Calculate $R_{m+1}(q_z)$ from $\varrho_{m+1}(z)$.
- (vii) If $|R(q_z) - R_{m+1}(q_z)| < \varepsilon$ for all q_z, then STOP! Take $\varrho(z) \equiv \varrho_{m+1}(z)$.
- (viii) Otherwise, take $\varrho_{m+1}(z)$ as new "guessed" profile $\varrho_m(z)$ and proceed from (iv).

It can be shown numerically that this algorithm is quite stable. Even "unknown layers", which were not put into the initial "guess", have been reconstructed [99]. Typically five to ten iterations are sufficient for a solution. The number of iterations depends not only on the quality of the guessed profile but also on the unknown profile itself. In general, the "phase-guessing" algorithm does not necessarily converge and examples reveal that for certain systems even very good input profiles are difficult to invert. A quite similar algorithm, where the dynamical scattering is better incorporated and only the difference between the unknown profile $\varrho(z)$ and the guessed profile $\varrho_m(z)$ is inverted, may be found in Refs. [99, 276].

As an example, the series of hexane films which was already described in Sect. 3.2.2 will be discussed, where the particular solid/liquid interface profile shown in the inset of Fig. 3.22 was proposed. This profile was obtained by inverting the reflectivity data of the whole series of wetting layers (see Fig. 3.21) according to the "phase-guessing" algorithm.

The data for a film with thickness $d = 86$ Å (open circles) together with calculations (lines) corresponding to the profiles in the inset are presented in Fig. 4.5. The upper profile was calculated for the system Si/SiO_2/hexane with error-function-shaped interface profiles. The best fit with this initial guess is depicted by the upper dashed line. Although not too bad, the general shape of the measurement cannot be explained. In a second step, a tiny feature at the interface was introduced that slightly improved the result (see second profile and second curve from the top in Fig. 4.5). The third profile and third curve are the results of an inversion of the whole data set consisting of 13 hexane films, according to the algorithm of the phase-guessing method. This profile was shown in Fig. 3.22. An almost perfect explanation of the data is achieved (see solid lines in the figure).

However, the result still has to be discussed carefully with respect to its uniqueness and reliability. In this specific case two additional, strong constraints were introduced: (i) For all 13 films the same solid/liquid interface profile was assumed. This assumption appears to be quite reasonable[8] and enormously increases the number of data points which define this interface.

[8] All films were measured in situ with the same wafer as substrate.

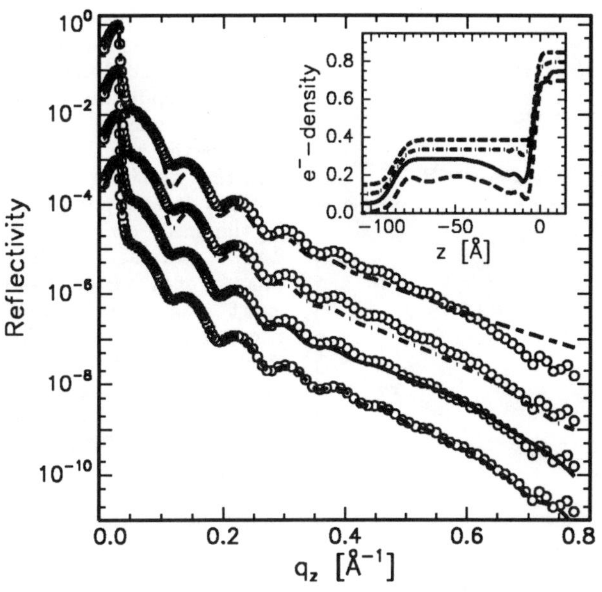

Fig. 4.5. Calculated reflectivities (*lines*) together with the data (*open circles*) for a hexane film of thickness $d = 86$ Å. The calculations were performed using the density profiles shown in the inset. The topmost profile corresponds to the topmost curve. In the second profile from the top a tiny feature at the solid/liquid interface was introduced. The lower two profiles were obtained with the phase-guessing method [96]

(ii) Parts of the profile were not allowed to be modified by the inverting algorithm. In fact, in the example only the solid/liquid interface region[9] was inverted while keeping all other interfaces at the values of the initial guess.

The fourth profile and curve in Fig. 4.5 are the result of an inversion where only the data from this particular film were taken and all regions of the profile were allowed to be altered. The calculated reflectivity perfectly coincides with the data. Compared to the third curve in Fig. 4.5 nothing really is gained, since data always have error bars and the obtained improvement is marginal. This shows that for this particular example a single reflectivity measurement would not be a sufficient basis for finding the real density profile.

A discussion of the phases corresponding to the respective density profiles yields more insight. Figure 4.6 depicts these phases for the starting profile of the phase-guessing method (see uppermost profile in the inset of Fig. 4.5) and the final result after inverting the data (corresponding to the third profile in the inset of Fig. 4.5). It can be seen that the two profiles have quite similar phases, which is also true for the respective Hilbert phases. Hence, this is the explanation of why the phase-guessing method works so well: All profiles that are not too different from the starting profile have very similar phases[10], i.e. they are mainly determined by the modulus of the structure factor, which is measured by x-ray reflectivity.

[9] The width of the liquid/gas interface and the mean layer thickness were two additional free parameters.

[10] They are not exactly equal, however, yielding e.g. the fourth profile in the inset of Fig. 4.5 as explained before.

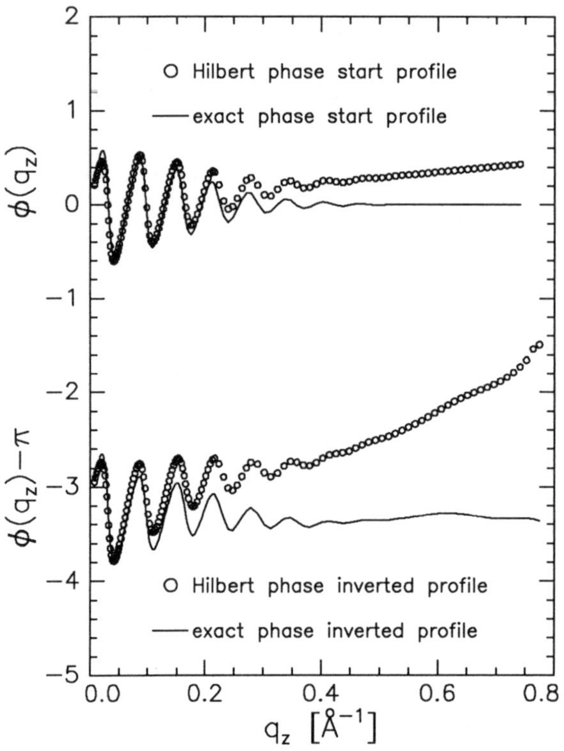

Fig. 4.6. *Upper curves*: Hilbert phase (*open circles*) and exact phase (*solid line*) corresponding to the electron density profile that serves as the start for the "phase-guessing" method (see upper profile in the inset of Fig. 4.5). *Lower curves*: Hilbert phase (*open circles*) and exact phase (*solid line*) corresponding to the electron density profile obtained after the inversion of the data with the "phase-guessing" method (see third profile in the inset of Fig. 4.5). Note that the average linear increase of the Hilbert phases is not real and may be suppressed by a redefinition of the $z = 0$ level

Finally, the following (trivial!) point has to be emphasized: Since it becomes possible to explain reflectivities almost perfectly with the "phase guessing" method one has to think carefully about possible systematic errors while the data are recorded. Very small features in a reflectivity may correspond to small but still detectable features of the profile!

4.3 Other Data Inversion Techniques

In the last section it was shown how a-priori information can be used to invert x-ray reflectivity data. However, the phase problem is not solved and ambiguities may still restrict the validity of the final result. Now other data inversion techniques are discussed, where the density profile of a system is reconstructed directly from reflectivity measurements without any specific knowledge about the sytem under investigation[11].

Recently, exact inversion schemes for neutron reflectivity data have been proposed [78, 79, 232, 233, 221]. They are essentially based on two basic

[11] Strictly speaking, this is not true, since at least the substrate material has to be known.

principles (i) If a known magnetic layer is added to an unknown system, the depth profile may be completely retrieved by performing three reflectivity measurements for different magnetizations and polarizations. (ii) The polarization of a neutron beam is used to obtain two independent reflectivities, thus allowing a back transformation of the data. Other proposed inversion methods are based on measurements with and without a reference layer [342] or a statistical Bayesian data analysis is proposed [343, 344].

Here, a special x-ray reflectivity inversion technique will be discussed: the use of anomalous x-ray scattering for reflectivity measurements. This method was first applied by *Sanyal et al.* [314, 336] to invert reflectivity data from a soft-matter thin film and later by *Ohkawa et al.* [268] for Al/C multilayers.

The principle of this method is quite simple. The actual system can be considered as a superposition of two systems. One system is assumed to be completely known, with an electron density $\varrho_0(z)$ and corresponding (complex) reflection coefficient $r_0(q_z)$ (given e.g. by the *Parratt* algorithm via Eq. 2.16). The unknown part is denoted by $\delta\varrho(z)$, and hence $\varrho(z) = \varrho_0(z) + \delta\varrho(z)$ is the total electron density profile of the sample. If $\delta\varrho(z)$ is small, the DWBA can be applied for the calculation of the reflectivity. The final result [314, 336] (r_e is the Thomson scattering length of the electron) is given by

$$R(q_z) = \left| \mathrm{i}\, r_0(q_z) + \frac{2\pi\, r_\mathrm{e}}{q_z} \left[a^2(q_z)\, \delta\tilde{\varrho}(q_{z,\mathrm{t}}) + b^2(q_z)\, \delta\tilde{\varrho}^*(q_{z,\mathrm{t}}) \right] \right|^2 . \qquad (4.15)$$

For the known profile $\varrho_0(z)$ a single layer with constant density ϱ_0 and thickness d was taken. The quantities $a(q_z)$ and $b(q_z)$ are the amplitudes of the transmitted and reflected waves, respectively, and $q_{z,\mathrm{t}}$ is the wavevector transfer inside this layer. These quantities, together with $r_0(q_z)$, may be easily calculated from the known density of the substrate and the formulas given in Sect. 2.2 (see Eqs. 2.16–2.20). Roughness may be included via *Névot–Croce* factors (see Eq. 2.34 in Sect. 2.3) obtained from previous measurements of the bare substrate. Furthermore, $\delta\tilde{\varrho}(q_z)$ is the Fourier transform of $\delta\varrho(z)$. If two reflectivities are measured, one at and the other slightly away from an absorption edge of the substrate, then Eq. (4.15) can be used to determine the real and imaginary parts of $\delta\tilde{\varrho}(q_{z,\mathrm{t}})$, and hence, after a Fourier back transformation, $\delta\varrho(z)$ may be directly calculated.

Figure 4.7a shows the density profile of a seven-layer CdA LB film on a Ge substrate (see Sect. 3.2.4) as obtained by inverting the data given by the symbols in Fig. 4.7b. Two reflectivities were taken at energies $E = 11\,103\,\mathrm{eV}$, close to the K edge of Ge and at $E = 10\,903\,\mathrm{eV}$, slightly away from the edge[12] (beamline X22C at the NSLS). This example proves that the method works in principle. Although this method does not have disadvantages such as the introduction of a new layer to the system, it is restricted to substrates

[12] The reflection coefficient of the substrate and its variation across the edge were measured in a subsidiary experiment.

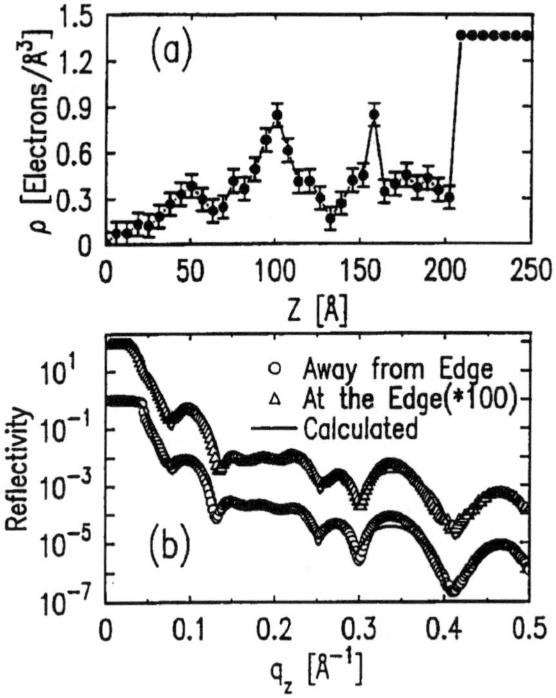

Fig. 4.7. **(a)** Electron density profile of a nominal seven layer Cd arachidate LB film as obtained by a model-independent inversion of anomalous reflectivity data. The *solid line* is a guide to the eye. The region $z > 202$ Å corresponds to the Ge substrate. **(b)** Reflectivity curves measured at and away from the K edge of the Ge substrate: *circles*, away from the edge; *triangles* at the edge ($\times 100$). The lines are calculations performed using the slicing method together with the *Parratt* algorithm and the density profile shown in Fig. 4.7a (from *Sanyal et al.* [314])

with absorption edges in the x-ray regime. Hence silicon, the most popular substrate material, is excluded.

The idea of anomalous x-ray reflectivity may also be formulated within the kinematical approximation, which should be used instead of the DWBA if data in a wide q_z range are inverted. The reason is that the DWBA may becomes inaccurate for too large q_z [335].

Again the unknown total density $\varrho(z) = \varrho_0(z) + \delta\varrho(z)$ is split into a completely known part $\varrho_0(z)$ and a part $\delta\varrho(z)$ that is to be determined. Thus, the structure factor defined by Eq. (4.1) may be written as

$$F(q_z) = F_0(q_z) + \delta F(q_z) = (X_0 + X) + \mathrm{i}\,(Y_0 + Y)\,, \tag{4.16}$$

where X_0, X and Y_0, Y denote the real and imaginary parts of $F_0(q_z)$ and $\delta F(q_z)$, corresponding to $\varrho_0(z)$ and $\delta\varrho(z)$, respectively. Two reflectivities $R^{(1)}$ and $R^{(2)}$ have to be measured, where $R^{(1)}$ is taken at an absorption edge and $R^{(2)}$ slightly away from an edge of the substrate, so that only X_0 and Y_0 change and X and Y are essentially the same. Then the unknown parts X and Y can be obtained by solving the equations[13]

[13] In Eqs. (4.17) and (4.18) the argument q_z of all quantities is omitted for reasons of clarity.

$$\frac{R^{(1)}}{R_{\mathrm{F}}^{(1)}} = (X_0^{(1)} + X)^2 + (Y_0^{(1)} + Y)^2 , \tag{4.17}$$

$$\frac{R^{(2)}}{R_{\mathrm{F}}^{(2)}} = (X_0^{(2)} + X)^2 + (Y_0^{(2)} + Y)^2 , \tag{4.18}$$

where $X_0^{(1,2)}, Y_0^{(1,2)}$, and $R_{\mathrm{F}}^{(1,2)}$ are the known substrate structure factors and Fresnel reflectivities for the two measurements. Finally, from X and Y, the density profile can be easily obtained by a Fourier transformation. However, the restriction of this method to certain rare substrates such as Ge, as mentioned before, is a disadvantage.

Often additional information about a system can be obtained from an investigation of the off-specular scattering. This is the topic of the next two chapters.

5. Statistical Description of Interfaces

In the previous chapters the lateral structure of interfaces was not considered, since reflectivity measurements are only sensitive to density profiles, i.e. to the laterally averaged structure. Before discussing off-specular scattering from interfaces in Chap. 6, a precise statistical description of surfaces has to be given.

Nowadays direct methods such as STM, AFM, and TEM are available and yield real-space images of surfaces or interfaces with atomic resolution[1]. This seemingly ultimate information has disadvantages, too. Often knowledge of the precise location of each single atom on a surface is not needed. In fact, the amount of recorded data may be huge compared to the information which is of interest. But another point is more important: Images obtained by microscopy yield *local* information over scales of several thousands of angstroms. X-ray scattering averages over much larger lateral length scales, which extend to several hundreds of microns in the regime of grazing angles [307, 369]. This means that x-ray scattering is in a sense complementary to the above-mentioned direct methods.

Hence, what is needed is a description of the surface morphology using only a few parameters corresponding to distinct features. Such a description is possible by using the language of statistics with correlation functions as the major element. For the precise statistical definitions a further study of the textbooks of *Bendat & Piersol* [32, 33], *Box & Jenkins* [45], or *Yaglom* [406] is suggested. The main focus in the next sections is on the application of correlation functions for the statistical description of surfaces.

5.1 Correlation Functions

To begin, a single interface described by the contour function $z(\boldsymbol{r}_{\|})$ with the lateral vector $\boldsymbol{r}_{\|} = (x, y)$ is considered. The case of multiple interfaces is considered in Sect. 5.4. The coordinate system is defined in the same manner as in Chap. 2 (see Fig. 2.6), with z being the vertical direction, x the lateral

[1] However, one always has to bear in mind that resolution effects, caused e.g. by the finite size of the tip of an STM or AFM, may also produce artefacts in these images.

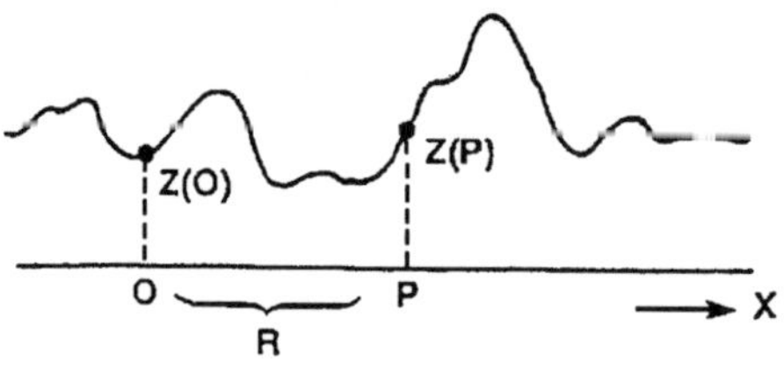

Fig. 5.1. Sketch of a surface contour $z(\boldsymbol{R})$. The height–height correlation function $C(\boldsymbol{R})$ yields the correlation of point $\mathcal{O}$ (origin) with point $\mathcal{P}$, spatially separated by $\boldsymbol{R}$ [335]

in-plane direction, and y the lateral out-of-plane direction. Further, it may be assumed that $z(\boldsymbol{r}_\parallel)$ has zero mean value, i.e. $\mu = \langle z(\boldsymbol{r}_\parallel)\rangle_{\mathcal{A}} = 0$, where $\mathcal{A}$ denotes a large spatial area[2]. It will be assumed for the moment that the extent of this area, over which all averages are performed in the following, is much larger than the characteristic lateral length scales of the roughness (e.g. the lateral correlation lengths ξ, see below). Later, when capillary waves are considered, this reasonable assumption will be partially relaxed. On the other hand, it is clear that the statistical description is restricted to length scales larger than the atomic dimensions (~ 1 Å) of the interfaces.

The function $z(\boldsymbol{r}_\parallel)$ contains the ultimate information about the interface: For each lateral point $\boldsymbol{r}_\parallel = (x, y)$, the value $z(\boldsymbol{r}_\parallel)$ is the height of the surface with respect to the mean interface location. However, for practical reasons, it may be more useful to describe the main features of an interface by only a few parameters. One of these parameters is the rms roughness σ, which has already been introduced in Sect. 2.3 (see Eq. 2.23).

A simple way to describe a surface by its statistical properties is given by the height–height correlation function $C(\boldsymbol{R})$ (see Fig. 5.1) defined as[3]

$$C(\boldsymbol{R}) = \frac{1}{\mathcal{A}} \int_{\mathcal{A}} z(\boldsymbol{r}_\parallel)\, z(\boldsymbol{r}_\parallel + \boldsymbol{R})\, \mathrm{d}\boldsymbol{r}_\parallel = \Big\langle z(\boldsymbol{r}_\parallel)\, z(\boldsymbol{r}_\parallel + \boldsymbol{R}) \Big\rangle_{\boldsymbol{r}_\parallel} , \tag{5.1}$$

where $\boldsymbol{R} = (X, Y)$ is a lateral vector and the $\boldsymbol{r}_\parallel$ average is taken over the large area $\mathcal{A}$. The definition given by Eq. (5.1) directly yields

$$\sigma^2 = C(\boldsymbol{0}) = \Big\langle z^2(\boldsymbol{r}_\parallel) \Big\rangle_{\boldsymbol{r}_\parallel} , \tag{5.2}$$

where σ is the rms roughness of the surface. Here it is worthwhile to recapitulate some definitions given in Sect. 2.3. The rms roughness may also be defined using the vertical probability density $P(z)$ (see Eq. 2.23)[4]:

---

[2] In principle, averages over many interface configurations have to be considered. However, for ergodic interfaces the configurational average may be replaced by the average over the lateral coordinates. Only this class of interfaces will be considered in this chapter.

[3] In the literature the height–height correlation function is also known as the autocorrelation function.

[4] Similarly, for the mean value of the interface height, $\mu = \int zP(z)\mathrm{d}z = \langle z(\boldsymbol{r}_\parallel)\rangle_{\boldsymbol{r}_\parallel} = 0$.

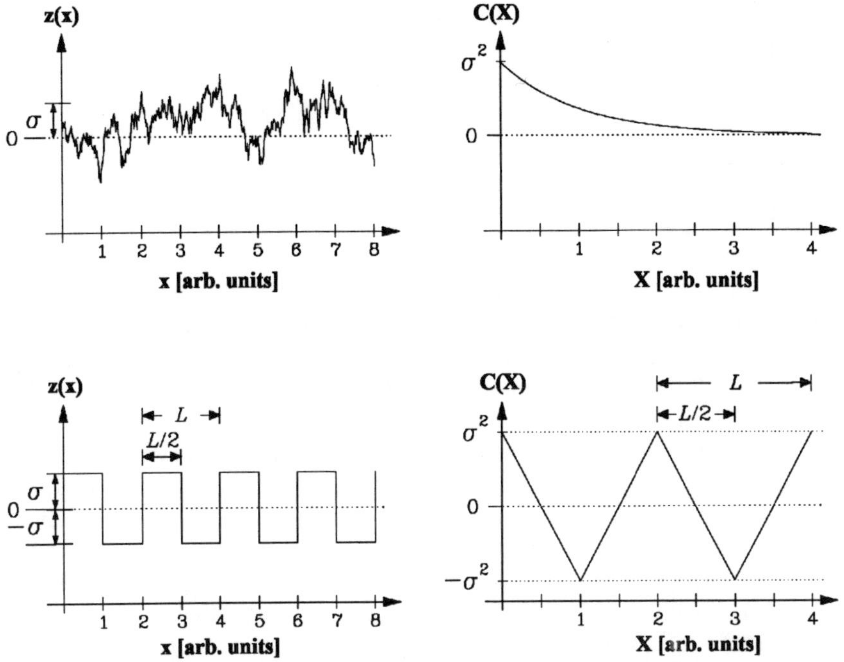

Fig. 5.2. Interface contour $z(x)$ and correlation function $C(X)$ of a statistically rough surface (*upper panels*) and a strictly periodic surface with periodicity L (*lower panels*). For clarity, only a one-dimensional description is given [354]

$$\sigma^2 = \int z^2 P(z)\,\mathrm{d}z = \frac{1}{\mathcal{A}} \iint\limits_{\mathcal{A}} z^2(x,y)\,\mathrm{d}x\,\mathrm{d}y \,. \tag{5.3}$$

Hence, lateral averages over the area $\mathcal{A}$ yield the same results as vertical averages over z with a certain probability density function $P(z)$. Therefore, a full statistical description of an interface needs the knowledge of both, a correlation function $C(\boldsymbol{R})$ for the lateral structure and a height distribution function $P(z)$ yielding the vertical density profile via integration [38] (see Sect. 2.3).

Before discussing particular correlation functions, some general aspects and a transformation to reciprocal space will be considered. If $P(z)$ is a Gaussian then all higher moments of the height distribution of the surface are fixed by $C(\boldsymbol{R})$. In this case the height–height correlation function provides a complete statistical description, showing its importance. The correlation function $C(\boldsymbol{R})$ is related to the probability of finding a point $\mathcal{P}$ at position $\boldsymbol{R}$ with the same height $z(\mathcal{P})$ above the mean interface as at the origin $\mathcal{O}$ (see Fig. 5.1). Figure 5.2 shows the correlation functions for two one-dimensional examples: a statistically rough surface and, for comparison, a periodic interface.

Another important quantity is the mean quadratic height difference function $g(\boldsymbol{R})$ defined by

$$g(\boldsymbol{R}) = \left\langle \left[z(\boldsymbol{r}_\parallel) - z(\boldsymbol{r}_\parallel + \boldsymbol{R})\right]^2 \right\rangle_{\boldsymbol{r}_\parallel} = 2\sigma^2 - 2C(\boldsymbol{R}) \,, \tag{5.4}$$

which is equivalent to

$$\frac{C(\boldsymbol{R})}{C(\boldsymbol{0})} = 1 - \frac{g(\boldsymbol{R})}{2\sigma^2} \,. \tag{5.5}$$

Equation (5.5) guarantees that $C(\boldsymbol{R}) \leq C(\boldsymbol{0}) = \sigma^2$ since always $g(\boldsymbol{R}) \geq 0$. On the other hand, if two points are uncorrelated, then $g(\boldsymbol{R}) = 2\sigma^2$, and $C(\boldsymbol{R}) = 0$ results. For a statistically rough surface an intermediate behavior is expected: $C(\boldsymbol{R})$ decreases monotonically until all correlations vanish for very large distances, i.e. $C(\boldsymbol{R}) \to 0$ for $|\boldsymbol{R}| \to \infty$.

In Fig. 5.2 the case of an ideally periodic surface is also shown. It is simple to prove that if the real-space function is periodic, i.e. $z(\boldsymbol{r}_\parallel) = z(\boldsymbol{r}_\parallel + m\boldsymbol{L})$ (m integer, $\boldsymbol{L}$ lateral vector), then the same holds for the correlation function, and thus $C(\boldsymbol{0}) = C(m\boldsymbol{L}) = \sigma^2$. If additionally $z(\boldsymbol{r}_\parallel) = -z(\boldsymbol{r}_\parallel + [m+1/2]\boldsymbol{L})$, then $C([m+1/2]\boldsymbol{L}) = -\sigma^2$. This means perfect anti correlation. Hence, the values of the height–height correlation function are restricted to the interval $-\sigma^2 \leq C(\boldsymbol{R}) \leq \sigma^2$ [354].

Furthermore, a simple calculation yields the following useful identity [38]:

$$\left\langle \left| \nabla_{\boldsymbol{r}_\parallel} z(\boldsymbol{r}_\parallel) \right|^2 \right\rangle_{\boldsymbol{r}_\parallel} = -\nabla^2_{\boldsymbol{R}} C(\boldsymbol{R}) \Big|_{\boldsymbol{R}=\boldsymbol{0}} \,. \tag{5.6}$$

Thus, the mean gradient of the surface is determined by the curvature of the correlation function at the origin [25]. This result is quite interesting since it allows a simple characterization of interfaces. For example, if a Gaussian $C(\boldsymbol{R}) = \sigma^2 \exp(-R^2/\xi^2)$ is taken, then $\langle |\nabla_{\boldsymbol{r}_\parallel} z(\boldsymbol{r}_\parallel)|^2 \rangle = 2\sigma^2/\xi^2$, and the interface appears laterally smooth for large lengths ξ. On the other hand, if an exponential $C(\boldsymbol{R}) = \sigma^2 \exp(-R/\xi)$ is assumed, then the second derivative diverges at $\boldsymbol{R} = \boldsymbol{0}$, and thus $\langle |\nabla_{\boldsymbol{r}_\parallel} z(\boldsymbol{r}_\parallel)|^2 \rangle = \infty$. It can be shown that all higher derivatives diverge, too [302]. Hence, this correlation function describes a jagged surface. The above examples will be discussed later in more detail (see Sect. 5.3).

5.2 Transformation to Reciprocal Space

A description of interfaces in reciprocal space is more convenient. The Fourier transform of the height function $z(\boldsymbol{r}_\parallel)$ is given by

$$\tilde{z}(\boldsymbol{q}_\parallel) = \int z(\boldsymbol{r}_\parallel) \exp(-\mathrm{i}\,\boldsymbol{q}_\parallel \cdot \boldsymbol{r}_\parallel)\, \mathrm{d}\boldsymbol{r}_\parallel = \left|\tilde{z}(\boldsymbol{q}_\parallel)\right| \exp\left\{\mathrm{i}\,\varphi(\boldsymbol{q}_\parallel)\right\} \,. \tag{5.7}$$

In the following the tilde always denotes Fourier-transformed quantities and $\boldsymbol{q}_\parallel = (q_x, q_y)$ is a lateral wavevector. The right-hand side of Eq. (5.7) is already separated into the modulus $|\tilde{z}(\boldsymbol{q}_\parallel)|$ and phase $\varphi(\boldsymbol{q}_\parallel)$. The so-called power spectral density (PSD)[5]

$$\tilde{C}(\boldsymbol{q}_\parallel) = \left|\tilde{z}(\boldsymbol{q}_\parallel)\right|^2 \tag{5.8}$$

is directly proportional to the scattered intensity only if very small wavevector transfers are involved ($|\boldsymbol{q}| < 0.01\,\text{Å}^{-1}$). This is true for light-scattering experiments[6] and also for soft-x-ray-scattering measurements ($\lambda > 50\,\text{Å}$) [37, 38, 41]. As will be shown in Chap. 6, for hard x-rays ($\lambda \sim 1\,\text{Å}$) the scattered intensity is *not* proportional to the PSD.

The definition shows that the PSD is a positive real quantity. The statistical properties of a surface are determined by the Fourier amplitudes $|\tilde{z}(\boldsymbol{q}_\parallel)|$ alone. The particular realizations $z(\boldsymbol{r}_\parallel)$ differ only in their respective phases $\varphi(\boldsymbol{q}_\parallel)$. The Wiener–Khinchin theorem guarantees that [406]

$$\tilde{C}(\boldsymbol{q}_\parallel) = \int C(\boldsymbol{R}) \exp(-\mathrm{i}\,\boldsymbol{q}_\parallel \cdot \boldsymbol{R})\,\mathrm{d}\boldsymbol{R}\,, \tag{5.9}$$

i.e. the PSD simply is the Fourier transform of the height–height correlation function. The free choice of the phases $\varphi(\boldsymbol{q}_\parallel)$ in reciprocal space corresponds to the fact that the correlation function in real space depends only on the spatial separation $\boldsymbol{R}$ of two points and not on their absolute locations. In other words; the statistical description of a surface in real space corresponds to the loss of phase information in reciprocal space. From Eq. (5.9), the useful identity

$$C(\boldsymbol{0}) = \sigma^2 = \int \tilde{C}(\boldsymbol{q}_\parallel)\,\mathrm{d}\boldsymbol{q}_\parallel \tag{5.10}$$

follows. This equation is of importance for the calculation of the rms roughness of soft-matter surfaces because $C(\boldsymbol{0})$ diverges for capillary waves (see Sect. 5.3.3). Since in practice only a finite $\boldsymbol{q}_\parallel$ range is accessible in a scattering experiment, Eq. (5.10) has to be replaced by

$$\sigma^2 = \int_{\boldsymbol{q}_\mathrm{min}}^{\boldsymbol{q}_\mathrm{max}} \tilde{C}(\boldsymbol{q}_\parallel)\,\mathrm{d}\boldsymbol{q}_\parallel\,, \tag{5.11}$$

with $\boldsymbol{q}_\mathrm{min}$ and $\boldsymbol{q}_\mathrm{max}$ defining the accessible region in reciprocal space. Equation (5.11) proves that the measured rms roughness, in principle, is a function of the experimental setup which determines $\boldsymbol{q}_\mathrm{min}$ and $\boldsymbol{q}_\mathrm{max}$ (see Sect. 3.1). This statement is also true for STM and AFM investigations where q_min is

[5] In principle, configurational averages over all surface realizations given by different phases $\varphi(\boldsymbol{q}_\parallel)$ have to be considered.

[6] For an overview in connection with liquid films see *Langevin* [203].

given by the inverse size of the sampled area and $q_{\max}$ by the inverse size of the tip.

If a correlation function $C(\boldsymbol{R})$ is given by a certain model, then a possible realization of a surface contour with the corresponding statistical properties may be obtained in the following way (Fourier-filter method [25]): With a Fourier transformation (see Eq. 5.9) one goes to reciprocal space, where, from Eq. (5.8), the amplitudes of $\tilde{z}(\boldsymbol{q}_\parallel)$ may be easily obtained. Then a phase is generated by $\varphi(\boldsymbol{q}_\parallel) = \mathrm{rnd}(\boldsymbol{q}_\parallel)$, where $\mathrm{rnd}(\boldsymbol{q}_\parallel)$ means a random number in the interval $[0, 2\pi]$. Finally, $\tilde{z}(\boldsymbol{q}_\parallel) = \sqrt{\tilde{C}(\boldsymbol{q}_\parallel)} \exp\{\mathrm{i}\,\mathrm{rnd}(\boldsymbol{q}_\parallel)\}$ has to be transformed back, yielding $z(\boldsymbol{r}_\parallel)$, i.e. a particular realization of the surface morphology. Examples of this procedure are given by *Barnsley et al.* [25], *Chiarello et al.* [59], and *Stettner* [323, 354] (see next section).

5.3 Some Examples

Examples of correlation functions will be discussed in this section. Since capillary waves are expected on most soft-matter thin-film surfaces, this topic will be explained in some more detail in Sect. 5.3.3.

5.3.1 Self-Affine Surfaces

Many structures in nature such as clouds, mountains, and coastlines, look similar on different length scales. They share the property of being fractals. However, this is not quite correct, since the vertical direction often scales differently compared to the lateral dimensions. Such a behavior is called *self-affine*. A nice macroscopic example of self-affine interfaces was given by *Engøy et al.* [112], who investigated simple cracks in wood. Those cracks possess self-affine rough interfaces with a roughness exponent $h = 0.7$ (see below).

In the past, surfaces were also described by self-affine models on a microscopic level. Various growth models yield this particular type of interface structure, where the KPZ model has become most prominent [182, 189, 190, 197, 227]. For example, under certain circumstances interfaces prepared by MBE lead to surfaces which have self-affine character [187, 271]. Details of fractal concepts in surface growth may be found in the book by *Barabási & Stanley* [23]. The interested reader is also referred to the recent review by *Krug* [198].

Many isotropic solid surfaces can be described by a simple correlation function which was proposed by *Sinha et al.* [335],

$$C(R) = \sigma^2 \exp\left\{ - (R/\xi)^{2h} \right\}, \tag{5.12}$$

with the cutoff (correlation) length ξ and the Hurst parameter h. The quantity ξ is the lateral length scale on which the interface begins to look rough:

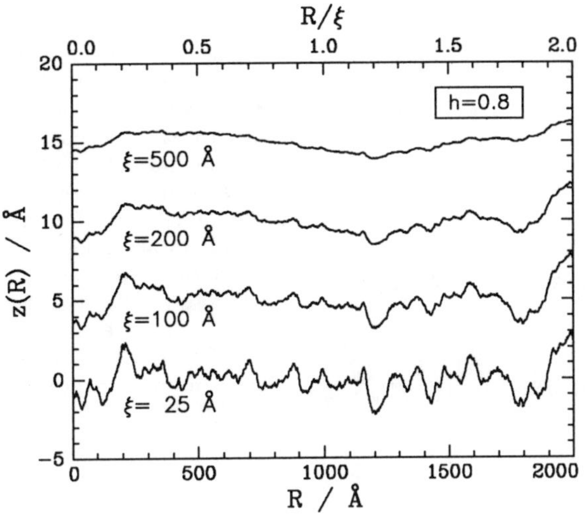

Fig. 5.3. Realizations of surfaces $z(R)$ corresponding to a self-affine correlation function $C(R) = \sigma^2 \exp\{-(R/\xi)^{2h}\}$ as calculated with the Fourier-filter method. The rms roughness for all surfaces is $\sigma = 1$ Å and the Hurst parameter $h = 0.8$. It can be seen that the cut-off length ξ controls the size of typical lateral distances [323, 354]

For $R \ll \xi$ the surface is self-affine rough, whereas for $R \gg \xi$ the surface appears to be smooth. The Hurst parameter h is restricted to the range $0 < h \leq 1$ and defines the fractal box dimension $D = 3 - h$ of the interface [235]. Small values of h correspond to extremely jagged surfaces, while values close to unity lead to interfaces with smooth hills and valleys. This may also be proved analytically by applying the criterion given by Eq. (5.6)[7].

Figures 5.3 and 5.4 visualize the meanings of these parameters. The interface contours were calculated with the Fourier-filter method described in the last section, for different ξ and h values and the same rms roughness $\sigma = 1$ Å.

Equation (5.12) takes into account the fact that the height difference function $g(R)$ defined by Eq. (5.4) has to satisfy $g(R) \sim R^{2h}$ for $R \ll \xi$, which is characteristic of self-affine interfaces[8]. On the other hand, Eq. (5.12) also causes $g(R)$ to saturate at $g(R) = 2\sigma^2$ for R values much larger than the length ξ. Otherwise the quadratic height difference would diverge for $R \to \infty$ [23, 335].

Although Figs. 5.3 and 5.4 reveal that various kinds of surfaces may be described by just three parameters σ, ξ, and h, there is one shortcoming if Eq. (5.12) is used as the correlation function: The limit $h \to 0$ is not defined properly. One would expect from growth models a correlation function $C(R) \sim \ln(R/\xi)$ (Edwards–Wilkinson growth [108]), whereas Eq. (5.12)

[7] Calculations by *Majaniemi, Ala-Nissila & Krug* [231] reveal that Eq. (5.12) does not show the correct limit for large R. Here the more general form $C(R) \sim \sigma^2 (R/\xi)^{-\gamma} \exp\{-c\,(R/\xi)^{2\hat{h}}\}$, with certain exponents γ and $\hat{h}$, and a (complex!) constant c is suggested.

[8] This is the reason why h often is called the roughness exponent.

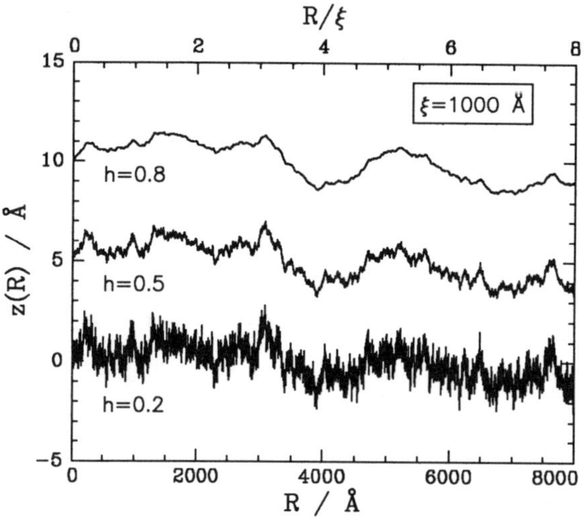

Fig. 5.4. Realizations as in Fig. 5.3, but for fixed $\xi = 1000\,\text{Å}$ and different h values. For small h the interface becomes extremely jagged, while it is smooth for larger h. The value $h = 0.5$ corresponds to an exponential correlation function and $h = 1$ to a Gaussian. For $h < 1$ the mean gradient $\langle|\nabla z(\boldsymbol{r}_\parallel)|^2\rangle$ of the surface contour diverges (see Eq. 5.6) [323, 354]

yields the constant $C(R) = \sigma^2/e$. This problem is solved in the next subsection with a detour to Fourier space.

5.3.2 *K*-Correlation Functions

Palasantzas [271, 272, 273] worked intensively on height–height correlation functions and found an interesting class, the so-called K-correlation functions, with the correct logarithmic limit for $h \to 0$.

To show how these K-correlation functions are obtained, the Fourier transform of an exponential, which corresponds to $h = 1/2$ in Eq. (5.12), is considered. The respective PSD is given by

$$\tilde{C}(\boldsymbol{q}_\parallel) = \frac{\sigma^2\,\xi^2}{(1+\boldsymbol{q}_\parallel^2\,\xi^2)^{3/2}}\,. \tag{5.13}$$

This is generalized by introducing $1+h$ in the denominator instead of $1+1/2 = 3/2$. To ensure the self-affine nature of the correlation function, i.e. $g(R) \sim R^{2h}$ for $h \neq 0$ and $g(R) \sim \ln(R)$ for $h = 0$, and to fulfill the normalization condition $\sigma^2 = \int \tilde{C}(\boldsymbol{q}_\parallel)\mathrm{d}\boldsymbol{q}_\parallel$, a dimensionless parameter b is introduced in the denominator. Thus, the PSD

$$\tilde{C}(\boldsymbol{q}_\parallel) = \frac{\sigma^2\,\xi^2}{(1+b\,\boldsymbol{q}_\parallel^2\,\xi^2)^{1+h}} \tag{5.14}$$

is proposed, where b is implicitly given by[9]

$$b = \frac{1}{2h}\left\{1-(1+b\,q_c^2\xi^2)^{-h}\right\}, \tag{5.15}$$

[9] For $h = 0$ the limit $h \to 0$ has to be taken.

with $q_c = \pi/\kappa$ being an upper wavevector cutoff fixed by the size of the molecules κ. A Fourier back transformation yields the real-space correlation function [272]

$$C(R) = \frac{\sigma^2}{b\,\Gamma(1+h)} \left(\frac{R}{2\,\xi\sqrt{b}}\right)^h K_h\left(\frac{R}{\xi\sqrt{b}}\right) , \tag{5.16}$$

where $K_h(x)$ denotes the modified Bessel function of the second kind[10] and of order h, and $\Gamma(x)$ denotes the Gamma function [4, 147]. The correlation function defined by Eq. (5.16) yields the correct limit $C(R) \to \ln(R)$ for $h \to 0$ and $R \ll \xi$, since $K_0(R) \sim \ln(R)$ for small arguments [4, 147]. For $h = 1/2$ the form given by Eq. (5.12) is regained and all other h values of the interval $0 \le h \le 1$ yield a self-affine correlation function $C(R)$. However, since Eq. (5.16) is a little complicated, Eq. (5.12) often is preferred[11]: $C(R)$ given by Eq. (5.12) can be computed much faster than Eq. (5.16). This is of great importance for the calculation of the final scattering formula (see Chap. 6) and a subsequent inclusion into χ^2 algorithms for fitting.

A distinct feature of the PSDs given by Eq. (5.14) is that they describe isotropic surfaces. But many surfaces grown by MBE or similar techniques are asymmetric. Steps begin to form at vicinal interfaces because of a small misorientation angle between the interface and a crystallographic plane. These steps lead to anisotropic surfaces and may be identified by scattering experiments that illuminate the sample parallel and perpendicular to these steps. This scenario was described quantitatively using statistical methods by *Pukite et al.* [289], who tried to explain RHEED experiments. Recently, *Holý et al.* [166] and *Kondrashkina, Stepanov, Opitz et al.* [193] transferred this formalism to the case of x-ray scattering. In the latter work the following PSD for a randomly stepped surface was found:

$$\tilde{C}(q_x) = \frac{2\,\xi_L\,\sigma_{\text{eff}}^2\,q_z^2}{(q_x\xi_L + q_z\theta_{\text{m}}\xi_L)^2 + q_z^4\,\sigma_{\text{eff}}^4} , \tag{5.17}$$

where θ_{m} denotes the small miscut angle between the interface and a crystallographic plane, q_x is the parallel[12] and q_z the perpendicular wavevector transfer, $\xi_L = \langle L \rangle$ is the mean width of the steps of height $H = \theta_{\text{m}}\xi_L$, which are distributed according to a probability density $P(L) = \xi_L^{-1}\exp(-L/\xi_L)$ ("geometric staircase"), and $\sigma_{\text{eff}}^2 = \sigma^2 + \theta_{\text{m}}^2\xi_L^2/2$ with σ being the total rms roughness. The denominator of Eq. (5.17) contains the asymmetry since different values are obtained for q_x and $-q_x$. It turns out that x-ray scattering data from MOVPE-grown GaAs/$Ga_{1-x}In_x$As/GaPAs and MBE-grown AlAs/GaAs multilayers ("step flow mode") can be quantitatively explained

[10] This is the reason for the name K-correlation function.

[11] This is justified for $h > 0.2$.

[12] Only a one-dimensional description is given here. This is sufficient to explain most scattering experiments where the slits are wide open in the y direction (see Sect. 3.1.1 and Chap. 6).

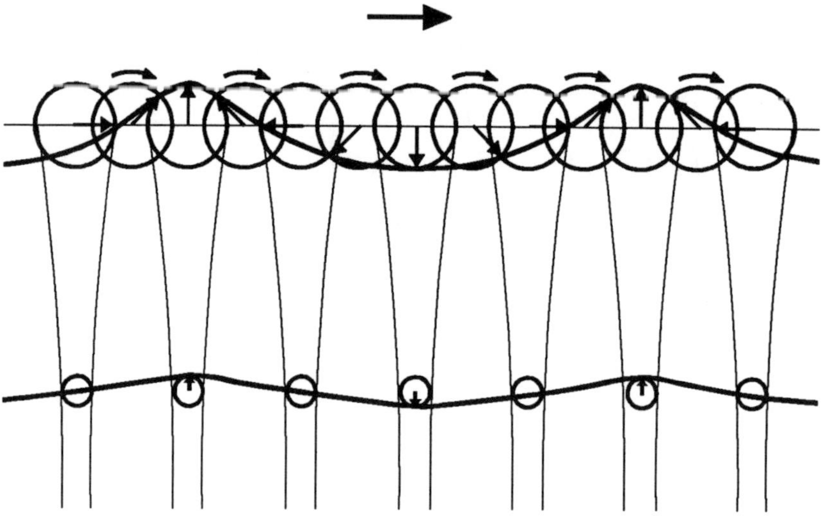

Fig. 5.5. Sketch of the particle motion in free capillary waves. The circular orbits are shown. The amplitude of the orbits decreases exponentially with increasing depth. In the case of a thin film the motion is squeezed vertically, leading to ellipsoidal orbits. This simple picture is only valid in the absence of viscosity [137]

assuming this asymmetric correlation function [166, 193]. Finally, two further points should be noted: (i) The correlation function given by Eq. (5.17) corresponds to a self-affine surface with $h = 1/2$ if $q_x \xi \ll 1$ [193, 337, 338]. (ii) *Stettner et al.* [287, 354, 355] performed extensive numerical calculations on step distributions, showing that even for small variations of the step size a description with the self-affine model (Eq. 5.12) is possible and that vertically a Gaussian probability density $P(z)$ can indeed be assumed. This is of fundamental importance for the diffuse scattering since most scattering theories implicitly contain this assumption (see Chap. 6).

5.3.3 Capillary Waves and Polymer Surfaces

Wave excitations are always present on liquid surfaces. If the effect of gravity is much larger than that of the surface tension, these waves are called *gravity waves*. Otherwise so-called *capillary waves* are found. The latter are of particular importance for most soft-matter surfaces. A characteristic feature of capillary waves is that their amplitude is much smaller than their wavelength.

The movement of the molecules in surface waves is schematically shown in Fig. 5.5. The velocity potential $\phi(x, z)$ of the particles[13] can be calculated from hydrodynamics with the result $\phi(x, z) = \phi_0 \exp(-qz) \cos[q_x x - \omega(q_x)t]$, i.e. waves of frequency ω, damped exponentially in the vertical z direction,

[13] The velocity of the particles is given by $\boldsymbol{v}(x, z) = \nabla\phi(x, z)$.

are found [203]. A liquid film that completely wets a solid substrate will now be considered more quantitatively.

The excess free energy per unit area of a film of thickness d and surface contour $z_l(\boldsymbol{r}_\parallel)$ on top of a substrate with the contour $z_s(\boldsymbol{r}_\parallel)$ is given by [14, 53, 298]

$$\Delta F\Big[z_s(\boldsymbol{r}_\parallel), z_l(\boldsymbol{r}_\parallel)\Big] = \int \Big\{(\gamma_{sf} - \gamma_{sv})\Big(\sqrt{1 + |\nabla z_s(\boldsymbol{r}_\parallel)|^2} - 1\Big) + P[z_l(\boldsymbol{r}_\parallel)] + \gamma\Big(\sqrt{1 + |\nabla z_l(\boldsymbol{r}_\parallel)|^2} - 1\Big) + \Delta\mu\Big[d + z_l(\boldsymbol{r}_\parallel)\Big]\Big\} \, \mathrm{d}\boldsymbol{r}_\parallel \,, \tag{5.18}$$

where γ is the surface tension of the liquid, γ_{sf} is the interfacial tension between the substrate and the film, and γ_{sv} that between the substrate and the vapor; $P[z_l(\boldsymbol{r}_\parallel)]$ is the deformation energy of the film surface with respect to a background potential and $\Delta\mu$ is the chemical-potential difference between the liquid and the vapor. Since only the situation near the vapor/liquid coexistence line will be discussed, $\Delta\mu = 0$ is assumed in the following. The square-root terms in Eq. (5.18) correspond to the excess deformation energies at the solid/liquid and liquid/vapor interfaces, respectively. The local film thickness at a lateral position $\boldsymbol{r}_\parallel$ is defined by $d(\boldsymbol{r}_\parallel) = d + z_l(\boldsymbol{r}_\parallel) - z_s(\boldsymbol{r}_\parallel)$, where $z_l(\boldsymbol{r}_\parallel)$ and $z_s(\boldsymbol{r}_\parallel)$ have zero mean values. It should be noted that Eq. (5.18) is only valid in the absence of viscosity. Figure 5.6 gives a picture of this situation.

First, the simplest case of a bulk-liquid surface without substrate will be considered. Then Eq. (5.18) can be simplified to [48, 53]

$$\Delta F\Big[z_l(\boldsymbol{r}_\parallel)\Big] \simeq \frac{\gamma}{2} \int \Big\{|\nabla z_l(\boldsymbol{r}_\parallel)|^2 + q_{l,c}^2 \, z_l^2(\boldsymbol{r}_\parallel)\Big\} \, \mathrm{d}\boldsymbol{r}_\parallel \,, \tag{5.19}$$

where $q_{l,c}$ is defined below, the square root term in Eq. (5.18) has been expanded assuming $|\nabla z_l(\boldsymbol{r}_\parallel)|^2 \ll 1$, and the background potential $P[z_l(\boldsymbol{r}_\parallel)] = g\varrho z_l^2(\boldsymbol{r}_\parallel)/2$ due to gravitation ($g = 9.81\,\mathrm{m/s^2}$ is the acceleration due to gravity and ϱ is the density of the liquid[14]) is used. Introducing the Fourier transform $\tilde{z}_l(\boldsymbol{q}_\parallel)$ of $z_l(\boldsymbol{r}_\parallel)$ leads to

$$\Delta F\Big[z_l(\boldsymbol{r}_\parallel)\Big] \simeq \frac{\gamma}{2} \int \Big(q_\parallel^2 + q_{l,c}^2\Big)\Big|\tilde{z}_l(\boldsymbol{q}_\parallel)\Big|^2 \, \mathrm{d}\boldsymbol{q}_\parallel \,. \tag{5.20}$$

The expression inside the integral is equivalent to the energy per oscillation mode with wavevector $\boldsymbol{q}_\parallel$. Taking into account the density of states (factor $4\pi^2$) and applying the equipartition theorem (energy per mode $= k_B T/2$) finally yields

$$\tilde{C}(\boldsymbol{q}_\parallel) = \Big|\tilde{z}_l(\boldsymbol{q}_\parallel)\Big|^2 = \frac{B}{4\pi} \frac{1}{q_\parallel^2 + q_{l,c}^2} \tag{5.21}$$

[14] Strictly speaking, one has to use the density difference between the liquid and the vapor.

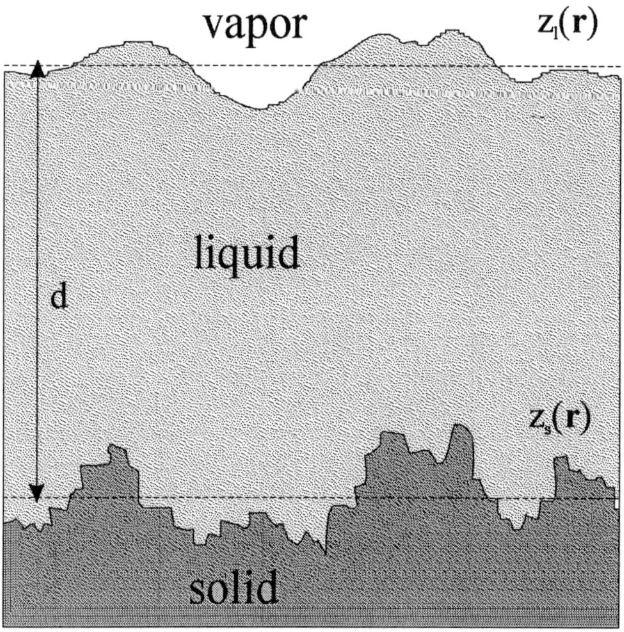

Fig. 5.6. Schematic drawing of a liquid film of mean thickness d on top of a rough substrate with contour $z_{\mathrm{s}}(\boldsymbol{r}_{\|})$. The surface of the liquid is given by $z_{\mathrm{l}}(\boldsymbol{r}_{\|})$. The contour $z_{\mathrm{s}}(\boldsymbol{r}_{\|})$ is partially replicated on top of the liquid by $z_{\mathrm{l}}(\boldsymbol{r}_{\|})$ (see Sect. 5.4). The main contributions to the excess free energy are the deformation energies at the solid/liquid and liquid/vapor interfaces

as the PSD for a bulk-liquid surface with thermally excited capillary waves. The constants $B = k_{\mathrm{B}}T/(\pi\gamma)$ and $q_{\mathrm{l,c}} = \sqrt{\varrho g/\gamma}$ are the same as those introduced in Sect. 3.1.1 (see Eqs. 3.10–3.12). In particular, $q_{\mathrm{l,c}}$ is the gravitational cutoff q_{g}, which is on the order of $q_{\mathrm{l,c}} = q_{\mathrm{g}} \sim 10^{-8}$–$10^{-7}\,\text{Å}^{-1}$, and hence very small. A similar consideration for the dispersion relation $\omega(q_{\|})$ of capillary waves yields [30]

$$\omega^2(q_{\|}) = g\, q_{\|} + \frac{\gamma}{\varrho}\, q_{\|}^3\,. \tag{5.22}$$

Equation (5.22) means that at high frequencies the capillary waves are restored by the surface tension and at low frequencies by gravity. Equations (5.21) and (5.22) give a complete description of the statics and dynamics of waves on liquid-like surfaces within the framework of classical hydrodynamics. Before discussing extensions of these simple results, the transformation to real space wil be done.

Fourier-transforming Eq. (5.21) yields the height–height correlation function [313]

$$C(R) = \frac{B}{2}\, K_0(q_{\mathrm{l,c}}R)\,, \tag{5.23}$$

where $K_0(x)$ is the modified Bessel function of the second kind and of order zero [4, 147]. Hence, the correlation function corresponding to free capillary waves equals the K-correlation function for $h = 0$ of the previous subsection.

Since $q_{\mathrm{l,c}}$ is very small if only gravity is present as the background potential, $K_0(q_{\mathrm{l,c}}R)$ in Eq. (5.23) may be expanded for all practical applications. Thus, capillary waves yield a logarithmic correlation function

$$C(R) \simeq -\frac{B}{2}\left[\ln(q_{\mathrm{l,c}}R/2) + \gamma_{\mathrm{E}}\right], \tag{5.24}$$

where γ_{E} is Euler's constant [4, 147].

Since $C(0)$ diverges if Eq. (5.23) or Eq. (5.24) is used, one has to calculate the rms roughness via Eq. (5.11) as an integral over the PSD. Hence,

$$\sigma^2 = \frac{B}{2}\int_0^{q_{\mathrm{u,c}}} \frac{q_\parallel}{q_\parallel^2 + q_{\mathrm{l,c}}^2}\,\mathrm{d}q_\parallel = \frac{1}{4}B\ln\left(\frac{q_{\mathrm{u,c}}^2 + q_{\mathrm{l,c}}^2}{q_{\mathrm{l,c}}^2}\right) \tag{5.25}$$

is the expression for the roughness of a liquid surface with upper cutoff $q_{\mathrm{u,c}} = 2\pi/\kappa$ as described in Sect. 3.1.1. Here the character of the wavevector $q_{\mathrm{l,c}}$ as "lower cutoff" becomes clear: It prevents, similarly to the upper cutoff $q_{\mathrm{u,c}}$ on atomic length scales, the divergence of the rms roughness on very large length scales. Equation (5.25) is the justification for using Eq. (3.11) in Chap. 3 as the roughness contribution of capillary waves[15]. In real space the upper cutoff can also be introduced by using

$$C(R) = \frac{B}{2}K_0\left(q_{\mathrm{l,c}}\sqrt{R^2 + \kappa^2}\right), \tag{5.26}$$

instead of Eq. (5.23), as the correlation function [239].

The first extension of these results will be made to liquids of finite depth. A film of thickness d on a flat substrate, i.e. $z_{\mathrm{s}}(\boldsymbol{r}_\parallel) = 0$, interacting via long-range van der Waals interactions $w(r) = -A_{\mathrm{H}}/(\pi^2 r^6)$, is considered[16]. The strength of the substrate/film interaction is determined by the effective Hamaker constant A_{eff} of the system, which may be calculated from the Hamaker constants A_{H} of the individual components [170, 365, 387]. Now the background potential $P[z_{\mathrm{l}}(\boldsymbol{r}_\parallel)]$ in Eq. (5.18) is given by a gravitational term plus a van der Waals contribution[17] [76]:

$$P[z_{\mathrm{l}}(\boldsymbol{r}_\parallel)] = \frac{1}{2}\,g\,\varrho\,z_{\mathrm{l}}^2(\boldsymbol{r}_\parallel) + \frac{A_{\mathrm{eff}}}{12\pi}\left(\frac{1}{[d + z_{\mathrm{l}}(\boldsymbol{r}_\parallel)]^2} - \frac{1}{d^2}\right). \tag{5.27}$$

If the contour function $z_{\mathrm{l}}(\boldsymbol{r}_\parallel)$ is much smaller than the film thickness d, Eq. (5.27) may be simplified to

$$P[z_{\mathrm{l}}(\boldsymbol{r}_\parallel)] \approx \frac{1}{2}\left(g\,\varrho + \frac{A_{\mathrm{eff}}}{2\pi\,d^4}\right) z_{\mathrm{l}}^2(\boldsymbol{r}_\parallel), \tag{5.28}$$

---

[15] In Eq. (5.25) the lower limit of integration was set to 0. In experiments, as shown in Sect. 3.1.1, the resolution has to be used instead.

[16] If retardation effects are important, $w(r) \sim 1/r^7$ rather than $w(r) \sim 1/r^6$ [76, 106].

[17] Because of the spatial integrations over the substrate and the film, the van der Waals contribution is proportional to $1/z^2$ instead of $1/r^6$.

where linear $z_l(\boldsymbol{r}_\parallel)$ terms have been omitted since they are canceled by the integration in Eq. (5.18). With the definition

$$q_{l,c}^2 = \frac{g\,\varrho}{\gamma} + \frac{A_{\rm eff}}{2\pi\,\gamma\,d^4} = q_{\rm g}^2 + q_{\rm vdW}^2(d)\ , \qquad (5.29)$$

Eq. (5.19) is regained, and all consequences are identical to those for bulk liquids. In particular, Eq. (3.22) in Sect. 3.2.1, which was used to explain the roughness curve $\sigma_{\rm PS}(d)$ of PS and that of liquid hexane films, is now justified. The lower cutoff $q_{l,c}$ is given by two quantities: the above-mentioned very small gravitational cutoff $q_{\rm g}$ and the van der Waals cutoff $q_{\rm vdW}(d) = a/d^2$ with the length $a = \sqrt{A_{\rm eff}/(2\pi\,\gamma)}$ (see Sect. 3.2.1 and Eq. 3.20). Since $a \sim$ 5–10 Å for most liquids and polymers, $q_{l,c} \approx q_{\rm vdW}(d)$ is a very good approximation for film thicknesses $d < 1000$ Å. For $d \to \infty$ the gravitational cutoff is still important. A further consequence of $q_{l,c} \approx q_{\rm vdW}(d)$ is that the logarithmic correlation function given by Eq. (5.24) is no longer valid and the full Bessel function expression of Eq. (5.23) has to be used. The influence of the constraint geometry on capillary waves on top of a liquid thin film in real space is sketched in Fig. 5.7. The inverse of the cutoff restricts the wavelength Λ of waves that can propagate on the surface to $\Lambda < \Lambda_{\rm max} = 2\pi/q_{l,c}$. Figures 5.7a and b schematically depict the $\Lambda_{\rm max} \sim d^2$ dependence. Figure 5.7c shows the case of a very thin film. If the film thickness is sufficiently small, conformal roughness is expected to dominate the surface morphology suppressing all capillary waves (see Sect. 5.4) [365].

The major difference between bulk liquids and liquid films is found for the dispersion relation. It changes to [30, 226, 301]

$$\omega^2(q_\parallel) = \left[\left(g + \frac{A_{\rm eff}}{2\pi\,\gamma\varrho\,d^4}\right)q_\parallel + \frac{\gamma}{\varrho}\,q_\parallel^3\right]\,\tanh(q_\parallel\,d)\ , \qquad (5.30)$$

because the additional constraint $(\partial\phi/\partial z)_{z=0} = 0$ for the motion of the particles at the substrate has to be fulfilled[18]. The particles moving on circular orbits with exponentially decaying radii in the case of a bulk liquid (see Fig. 5.5) are forced onto ellipsoidal orbits in case of a thin film. However, this has no influence on the free energy and hence on the PSD and surface roughness [99]. It should be noted that the dispersion relation $\omega(q_\parallel)$ cannot be measured directly with inelastic x-ray scattering since the energy resolution at even the best facilities is not good enough yet. An extensive review of the possibilities and applications of inelastic x-ray scattering is given in the book by *Burkel* [54], which is recommended to the interested reader.

Equation (5.29) accounts for the modification of the lower cutoff by the van der Waals forces between the substrate and the liquid film. For an arbitrary potential $P[z_l(\boldsymbol{r}_\parallel)]$ the cutoff may be calculated via [218]

$$q_{l,c}^2 = \frac{1}{\gamma}\,\frac{\partial^2 P(z_l)}{\partial z_l^2}\bigg|_{z_l=0}\ . \qquad (5.31)$$

[18] This simply means $v_z = 0$ at the substrate.

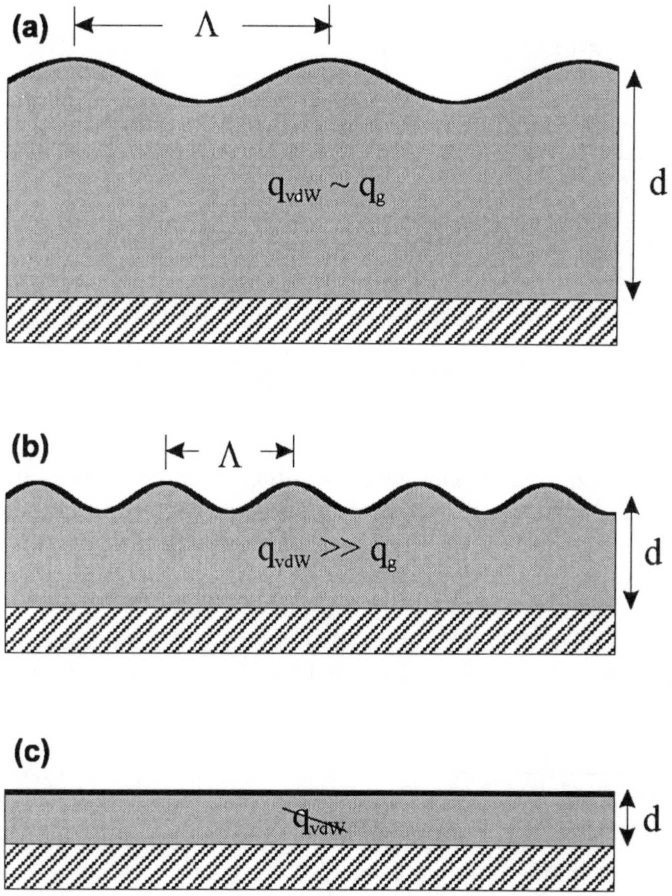

Fig. 5.7. Sketch of the influence of the constraint geometry on capillary waves. **(a)** For thick films the van der Waals cutoff $q_{\mathrm{vdW}}(d) \sim 1/d^2$ is expected to be on the order of the small gravitational cutoff q_{g}, thus allowing very long lateral wavelengths Λ. **(b)** For thinner films $q_{\mathrm{vdW}}(d)$ becomes larger and only waves with $\Lambda < 2\pi/q_{\mathrm{vdW}}(d) \sim d^2$ can propagate. **(c)** For a very thin film all capillary waves are suppressed by the substrate interaction

Several extensions of the above picture have been discussed. If, in Eq. (5.18), curvature corrections of the fluid interface are taken into account, the result for the PSD changes to [70, 95, 159, 160, 243]

$$\tilde{C}(\boldsymbol{q}_{\|}) = \frac{B}{4\pi} \frac{1}{(1+\bar{c}/\gamma)\, q_{\|}^2 + q_{\mathrm{l,c}}^2}, \tag{5.32}$$

where $\bar{c}$ can be calculated from the microscopic constants of the system (see *Napiórkowski & Dietrich* [95, 257, 258, 259]). The corresponding correlation function is given by

$$C(R) = \frac{B\gamma}{2(\gamma+\bar{c})} K_0\left(\sqrt{\frac{\gamma}{\gamma+\bar{c}}}\, q_{\mathrm{l,c}} R\right) . \tag{5.33}$$

Since $q_{\mathrm{l,c}} \sim 1/\gamma^{1/2}$ and $B \sim 1/\gamma$ the influence of curvature corrections on capillary waves may be taken into account by a replacement of the surface tension γ by an effective one $\gamma + \bar{c}$.

More subtle calculations by *Napiórkowski & Dietrich* [95, 257, 258, 259] taking account of the fact that the long-range microscopic pair potential may be better described by $w(r) \sim 1/(\kappa^2 + r^2)^3$ instead of $w(r) \sim 1/r^6$ yield the following expansion for $\tilde{C}^{-1}(\boldsymbol{q}_{\|})$:

$$\frac{B}{4\pi}\frac{1}{\tilde{C}(\boldsymbol{q}_{\|})} = q_{\mathrm{l,c}}^2 + q_{\|}^2 - \frac{1}{4} q_{\|}^4 \kappa^2 \Big|\ln(\kappa q_{\|})\Big| + O\Big[(q_{\|}\kappa)^6 \ln(q_{\|}\kappa)\Big] , \tag{5.34}$$

with a microscopic length[19] κ. The first two terms in Eq. (5.34) equal those of the denominator in Eq. (5.21). The next-to-leading asymptotic behavior is given by the term $q_{\|}^4\kappa^2|\ln(q_{\|}\kappa)|$ with a negative sign. This conceptually very important correction has not yet been observed in experiments. Furthermore, recent calculations by *Mecke & Dietrich* [241] also predict odd terms proportional to $q_{\|}^3$ on the right-hand side of Eq. (5.34).

If the square root in Eq. (5.18) is expanded to fourth order, then similar arguments to those which lead to Eq. (5.21) yield a $q_{\|}^4$ term in the denominator of the PSD [11, 65, 66, 67, 204, 240, 243], and thus

$$\tilde{C}(\boldsymbol{q}_{\|}) = \frac{B}{4\pi} \frac{1}{(K/\gamma)\, q_{\|}^4 + q_{\|}^2 + q_{\mathrm{l,c}}^2} . \tag{5.35}$$

Here $K = K_1 + K_2$ denotes the bending elasticity modulus of the liquid surface K_1 plus a mode–mode coupling term K_2. The latter is estimated as $K_2 \sim 3\,k_{\mathrm{B}}T/(8\pi)$ [66, 243]. The respective height–height correlation function is obtained by Fourier transforming Eq. (5.35)[20]:

$$C(R) = \frac{B}{2}\Big[K_0(q_{\mathrm{l,c}}R) - K_0(q_K R)\Big] , \tag{5.36}$$

with the additional cutoff $q_K = (\gamma/K)^{1/2}$. In Eq. (5.36) the limit $R \to 0$ exists and the rms roughness is simply given by

$$\sigma^2 = C(0) = \frac{B}{2}\ln\left(\frac{q_K}{q_{\mathrm{l,c}}}\right) . \tag{5.37}$$

Hence, q_K serves as an upper cutoff here. The bending elasticity K_1 is of importance for smectic and freestanding soft-matter thin films (see Refs. [68, 69, 211, 247, 248, 334, 381, 382]).

[19] All results are equivalent to a description with a $q_{\|}$-dependent surface tension [11, 95].

[20] For a detailed calculation see *McClain et al.* [240].

Fredrickson et al. [128] calculated the PSD for a molten polymer brush, such as a layer of diblock copolymer at an air interface, within the Alexander–de Gennes approximation [6, 75] of a delta-function distribution of chain ends. They obtained for a film of thickness d the following result:

$$\tilde{C}(\boldsymbol{q}_{\|}) = \frac{B}{4\pi} \frac{q_{\|}^2}{q_{\|}^4 + q_{\mathrm{F}} q_{\|}^3 + 3 q_{\mathrm{F}}\, d^{-3}/2} , \tag{5.38}$$

where $q_{\mathrm{F}} = 2\mu_0/\gamma$ is determined by the ratio of the bulk shear modulus μ_0 and the surface tension γ of the polymer. Capillary waves were not considered. The long-wavelength modes in a brush are suppressed because of the lateral chain stretching required to maintain constant density. A discussion of Eq. (5.38) reveals that the *Fredrickson* PSD has a broad maximum. The location q_{M} of this maximum plays, in a sense, the role of the lower cutoff $q_{\mathrm{l,c}}$ (see the discussion in Sect. 6.2.1). One obtains

$$q_{\mathrm{M}} = 3^{1/3}\, d^{-1} \qquad \text{for large } q_{\mathrm{F}} , \tag{5.39}$$

$$q_{\mathrm{M}} = (3 q_{\mathrm{F}}/2)^{1/4}\, d^{-3/4} \qquad \text{for small } q_{\mathrm{F}} , \tag{5.40}$$

as explicit expressions, which may be summarized as

$$q_{\mathrm{M}} \approx a/d^m \approx q_{\mathrm{l,c}} \quad \text{with } m \in [0.75, \ldots, 1] , \tag{5.41}$$

as already used in Sect. 3.2.1 for the explanation of the PS measurements. As before, the rms roughness is obtained by integrating Eq. (5.38) over all wavenumbers, with the result[21]

$$\sigma^2 = \frac{B}{2} \ln\left(q_{\mathrm{u,c}}\, d^m/a\right) , \tag{5.42}$$

yielding a logarithmic increase of the surface roughness with the film thickness (see Eq. 3.21).

In Sect. 3.2.1 an unusual roughness curve observed for a series of PEP films was described (see Fig. 3.13). The explanation given there was based on capillary waves which were superimposed on an island microstructure. The size of these islands $l(d)$ was assumed to depend on the film thickness as $l(d) \sim d$, defining a wavenumber $q_0 = 2\pi/l(d) \approx 1/d$. Hence, a simple ad-hoc assumption for the PSD would be

$$\tilde{C}(\boldsymbol{q}_{\|}) = \frac{B}{4\pi} \frac{1}{q_{\|}^2 + q_{\mathrm{l,c}}^2} \frac{q_0 + q_{\|}}{q_{\|}} . \tag{5.43}$$

Equation (5.43) is purely empirical and shows the correct limit for $q_{\|} \gg q_0$. It was integrated over $q_{\|}$ to obtain the theoretical roughness curve $\sigma_{\mathrm{PEP}}(d)$ shown in Fig. 3.13 (see Eq. 3.25).

Fourier transformation of Eq. (5.43) yields the height–height correlation function for a liquid surface with an island microstructure

[21] For simplicity the $q_{\|}^3$ term in Eq. (5.38) has been dropped and an upper cutoff $q_{\mathrm{u,c}}$ has been introduced.

Table 5.1. Summary of the cutoffs (third column) (see also Fig. 5.7) and expansions for large $q_\|$ of the different PSDs $\tilde{C}(\boldsymbol{q}_\|)$, as discussed in the text, corresponding to the equations given in the second column. For comparison the last line contains the corresponding properties of a K-correlation function that describes a self-affine surface with a Hurst parameter h and a lateral correlation length ξ (see Sect. 5.3.2)

$\tilde{C}(\boldsymbol{q}_\|)$	Equation	Cutoffs	Expansion
Free capillary waves	(5.21)	q_g	$q_\|^{-2}$
Thin film	(5.21), (5.29)	$q_{vdW}(d)$	$q_\|^{-2}$
Curvature corrections	(5.32)	$[\gamma/(\gamma+\bar{c})]^{1/2} q_{l,c}$	$q_\|^{-2}$
Modified potential	(5.34)	q_g, $q_{vdW}(d)$	$(q_\|^2 - q_\|^4 \vert \ln q_\| \vert)^{-1}$
Bending included	(5.35)	q_g, $q_{vdW}(d)$, q_K	$q_\|^{-4}$
Polymer brush	(5.38)	q_F, q_M	$q_\|^{-2}$
Waves + islands	(5.43), (5.46)	$q_{vdW}(d)$, q_0	$q_\|^{-3}$
K-correlation	(5.14)	$2\pi/(\xi\sqrt{b})$	$q_\|^{-2-2h}$

$$C(R) = \frac{B}{2}\left\{K_0(q_{l,c}R) + \frac{\pi\, q_0}{2\, q_{l,c}}\Big[I_0(q_{l,c}R) - L_0(q_{l,c}R)\Big]\right\}, \tag{5.44}$$

where $q_{l,c}$ is given by Eq. (5.29), $K_0(x)$ and $I_0(x)$ are modified Bessel functions, and $L_0(x)$ is a modified Struve function [4, 147]. We will come back to this correlation function in Chap. 6.

However, Eq. (5.44) does not take into account the fact that the islands on the PEP surface may form a rather regular arrangement. Since a strictly periodic surface would yield a periodic correlation function (see Fig. 5.2), and one may separate the displacement term $z(\boldsymbol{r}_\|)$ into a capillary wave and an island part, i.e. $z(\boldsymbol{r}_\|) = z_{cap}(\boldsymbol{r}_\|) + z_{is}(\boldsymbol{r}_\|)$, where the two fluctuations are independent, the real-space function [376]

$$C(R) = \frac{B}{2} K_0(q_{l,c}R) + h^2 \exp(-\Delta q\, R)\cos(q_0\, R) \tag{5.45}$$

is more reasonable. In Eq. (5.45), $q_0 = 2\pi/l(d)$ is the periodicity given by the average size $l(d)$ of the islands (see above), Δq defines the width of the respective size distribution in reciprocal space, and h is the island "rms roughness" (mean island height). It is obvious that both Δq and h are dependent on the island size and thus the film thickness (for details see Ref. [413]). The Fourier transform of Eq. (5.45) is

$$\tilde{C}(\boldsymbol{q}_\|) = \frac{B}{4\pi}\frac{1}{q_\|^2 + q_{l,c}^2} + \frac{h^2}{2\pi}\mathrm{Re}\left\{\frac{\Delta q - \mathrm{i}q_0}{\left[(\Delta q - \mathrm{i}q_0)^2 + q_\|^2\right]^{3/2}}\right\}, \tag{5.46}$$

where Re{...} means the real part of the complex quantity in the brackets. A consideration of Eq. (5.46) reveals that the PSD exhibits a peak for a narrow size distribution, or, more quantitatively, for $q_0 > \Delta q$. In the reverse case, i.e. if the mean value of the reciprocal island size q_0 is smaller than the width Δq of the respective distribution, the surface appears essentially random in all directions and the PSD does not show a peak corresponding to a preferred lateral length scale.

Table 5.1 summarizes the main properties of all PSDs for liquid and polymer interfaces presented in this subsection. In the last column the expansion for $q_\| \gg q_{l,c}$ is given. These expansions give a crude estimate of the expected diffusely scattered x-ray intensity for large $q_\|$ (see Sect. 6.1.3).

Finally, the influence of viscosity will be briefly discussed. *Jäckle* [173] calculated the dynamic susceptibility $\chi_{zz}(q_\|, \omega)$ for a liquid viscous film. In the limit of very thin films ($d \to 0$) and in the linear-response approximation the simple formula

$$\chi_{zz}(q_\|, \omega) = \frac{q_\|^2\, d}{-\varrho\,\omega^2 - 4\mathrm{i}\omega\,\eta\, q_\|^2 + \gamma\, q_\|^2 d\left(q_\|^2 + q_{l,c}^2\right)} \tag{5.47}$$

is obtained, where η denotes the viscosity. With the susceptibility $\chi_{zz}(q_\|, \omega)$ the dynamic PSD is given by [155, 172]

$$\tilde{C}(\boldsymbol{q}_\|, \omega) = \frac{2k_\mathrm{B}T}{\omega}\,\chi''_{zz}(q_\|, \omega)\ , \tag{5.48}$$

and the static PSD, which is important for x-ray scattering experiments, may be easily calculated via

$$\tilde{C}(\boldsymbol{q}_\|) = \tilde{C}(\boldsymbol{q}_\|, 0) = k_\mathrm{B}T\chi_{zz}(q_\|, 0)\ . \tag{5.49}$$

Thus, Eqs. (5.48) and (5.49) show that viscosity exactly cancels in a static description. Hence, the static correlation functions $C(R)$ of a liquid and of a glassy surface (i.e. a "frozen liquid") should be identical[22]. However, this result was derived only for the limit $d \to 0$. This problen is a current topic of research in the field of soft-matter thin films. The first reasonably complete results were recently presented by *Jäckle* [174].

5.4 Vertical Roughness Correlations

The description of an interface by one correlation function $C(R)$ is not complete if the interface contour is at least partially transferred from one layer to the next. This is called vertically correlated or conformal roughness. Figure 5.8a depicts this case schematically, and that of vertically uncorrelated interface roughness is shown in Fig. 5.8b.

The case of partially correlated interfaces will be considered in more detail. For clarity, no strict mathematical development is given. Here the reader is referred to the work of *Stettner et al.* [354, 355], where a mathematically

[22] Except for a change induced by the direct temperature dependence of $C(R)$.

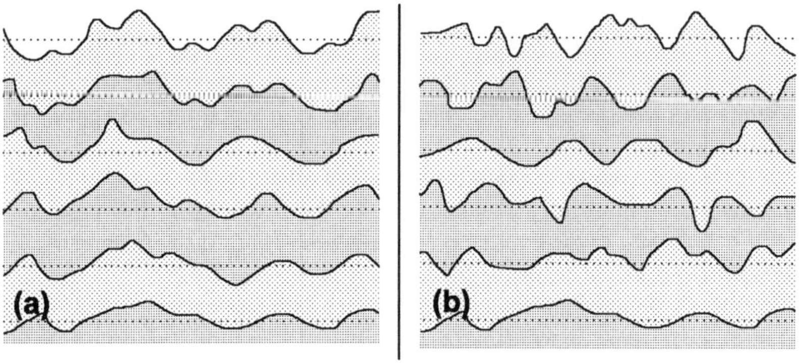

Fig. 5.8. Examples of multilayer systems. **(a)** Partially correlated interface roughness. The underlying interface structure is partially replicated by the overlayer. **(b)** Uncorrelated roughnesses. The interface contour of the overlayer is unaffected by the underlying structures. The *dotted lines* indicate the mean locations of the interfaces [354]

complete and exact description is presented. The same notation as introduced in Sect. 2.2 and Fig. 2.6 is used. The contour of interface j is denoted by $z_j(\boldsymbol{r}_{\|})$ and describes the deviation from the mean interface location μ_j at a lateral point $\boldsymbol{r}_{\|} = (x, y)$. Following *Spiller, Stearns & Krumrey* [346], the Fourier transform $\tilde{z}_k(\boldsymbol{q}_{\|})$ of interface k is given by a replicated part from interface j and an intrinsic part which would be present without the other interfaces. This may be expressed by

$$\tilde{z}_k(\boldsymbol{q}_{\|}) = \chi_{jk}(\boldsymbol{q}_{\|})\, \tilde{z}_j(\boldsymbol{q}_{\|}) + \tilde{z}_{k,\mathrm{intr.}}(\boldsymbol{q}_{\|}) \;, \tag{5.50}$$

where $\chi_{jk}(\boldsymbol{q}_{\|})$ is the so-called replication factor. This factor describes how the Fourier component of layer j with lateral wavevector $\boldsymbol{q}_{\|}$ is transferred to a layer k above. For $\chi_{jk}(\boldsymbol{q}_{\|}) = 1$, the component is perfectly replicated. The other extreme, $\chi_{jk}(\boldsymbol{q}_{\|}) = 0$, means a total suppression of this Fourier component. In Eq. (5.50) $\tilde{z}_{k,\mathrm{intr.}}(\boldsymbol{q}_{\|})$ denotes the intrinsic roughness that is unaffected by the other layers. In the case of fluid interfaces the intrinsic part would be given by capillary waves according to the descriptions in Sect. 5.3.3.

The physics of the system under consideration determines the precise expression for $\chi_{jk}(\boldsymbol{q}_{\|})$. However, only a few formulations of the replication factor based on microscopic models are available in the literature. These will be discussed later. First some more definitions have to be introduced. From Eq. (5.50) it follows directly that

$$\tilde{C}_{jk}(\boldsymbol{q}_{\|}) = \chi_{jk}(\boldsymbol{q}_{\|})\, \tilde{C}_j(\boldsymbol{q}_{\|}) \;, \tag{5.51}$$

where $\tilde{C}_j(\boldsymbol{q}_{\|}) = |\tilde{z}_j(\boldsymbol{q}_{\|})|^2$ is the PSD of interface j, and $\tilde{C}_{jk}(\boldsymbol{q}_{\|})$ is defined by[23]

[23] Note again that a stricter description would be based on configurational averages.

$$\tilde{C}_{jk}(\boldsymbol{q}_\parallel) = \left| \tilde{z}_j(\boldsymbol{q}_\parallel)\, \tilde{z}_k^*(\boldsymbol{q}_\parallel) \right| . \tag{5.52}$$

Equation (5.51) shows more clearly how the replication factor transfers the roughness in terms of the respective PSDs. Fourier-transforming Eq. (5.52) yields the real-space description (see Eq. 5.1)

$$C_{jk}(\boldsymbol{R}) = \frac{1}{\mathcal{A}} \int_{\mathcal{A}} z_j(\boldsymbol{r}_\parallel)\, z_k(\boldsymbol{r}_\parallel + \boldsymbol{R})\, \mathrm{d}\boldsymbol{r}_\parallel = \Big\langle z_j(\boldsymbol{r}_\parallel)\, z_k(\boldsymbol{r}_\parallel + \boldsymbol{R}) \Big\rangle_{\boldsymbol{r}_\parallel} . \tag{5.53}$$

The function $C_{jk}(\boldsymbol{R})$ is called the cross-correlation function[24]. The meaning of the correlation function $C_{jk}(\boldsymbol{R})$ is similar to that of the height–height correlation function of a single interface: The probability of finding a point on interface k at a position laterally separated by $\boldsymbol{R}$ from a point on interface j which has same height above the respective mean interface location is proportional to $C_{jk}(\boldsymbol{R})$. If roughness correlations are present then the height–height and cross-correlation functions are needed for a stochastically complete description of a layer system.

Hence, in principle roughness correlations can be described in real space without replication factors. However, it is not simple to find expressions for $C_{jk}(\boldsymbol{R})$. From the definition of $C_{jk}(\boldsymbol{R})$ two constraints follow,

$$|C_{jk}(\boldsymbol{R})|^2 \le C_j(\boldsymbol{0})\, C_k(\boldsymbol{0}) , \qquad |C_{jk}(\boldsymbol{R})| \le \frac{1}{2}\Big[C_j(\boldsymbol{0}) + C_k(\boldsymbol{0})\Big] , \tag{5.54}$$

which may help one to find reasonable expressions for cross-correlation functions [32, 33]. A simple function that is often used in the literature to describe the correlations between self-affine rough interfaces (see Sect. 5.3.1) is [244, 264, 269, 314, 337]

$$C_{jk}(R) = \sigma_j \sigma_k \exp\Big\{ -(R/\xi)^{2h} \Big\} \exp\Big\{ -|\mu_j - \mu_k|/\xi_{ik,\perp} \Big\} , \tag{5.55}$$

where σ_j, σ_k and μ_j, μ_k are the rms roughnesses and mean locations of interfaces j and k, respectively. The quantity $\xi_{ik,\perp}$ is a vertical correlation length. If $\xi_{ik,\perp}$ is much larger than the distance $|\mu_j - \mu_k|$ between the two layers, they are perfectly correlated. Although quite successful, Eq. (5.55) has one major shortcoming: All interfaces are assumed to have the same constant lateral correlation length ξ and Hurst parameter h. This means also that the roughness propagates equally for all lateral length scales. Since later, in Sect. 6.2.4, we will focus on the description in reciprocal space, no more details will be discussed here [354, 355].

Finally, the replication factor for a liquid-thin film surface, denoted in the following by $z_l(\boldsymbol{r}_\parallel)$, will be discussed (see Fig. 5.6). *Andelmann, Joanny & Robbins* [14, 298] calculated[25] $\chi(\boldsymbol{q}_\parallel)$ for the case of a single film of thickness d on top of a rough substrate with contour $z_s(\boldsymbol{r}_\parallel)$. By variation of the excess

[24] If $j = k$ then $C_{jj}(\boldsymbol{R})$ is equal to the height–height correlation function $C_j(\boldsymbol{R})$ of the interface $z_j(\boldsymbol{r}_\parallel)$ as defined by Eq. (5.1).

[25] The indices "jk" of the replication factor $\chi_{jk}(\boldsymbol{q}_\parallel)$ are omitted in the following.

free energy (see Eq. 5.18 in Sect. 5.3.3) with respect to $z_l(\boldsymbol{r}_\parallel)$, the following equation is obtained [14, 298]:

$$\gamma\nabla\cdot\left[\frac{\nabla z_l(\boldsymbol{r}_\parallel)}{\sqrt{1+|\nabla z_l(\boldsymbol{r}_\parallel)|^2}}\right]=\iint_{z_l(\boldsymbol{r}_\parallel)-z_s(\boldsymbol{r}_\parallel+\boldsymbol{r}'_\parallel)}^{\infty} w(\boldsymbol{r}'_\parallel,z')\,\mathrm{d}z'\,\mathrm{d}\boldsymbol{r}'_\parallel\,, \quad (5.56)$$

where $w(\boldsymbol{r}'_\parallel,z')$ expresses the interaction potential between the molecules of the liquid and the substrate, and $\boldsymbol{r}'_\parallel=(x',y')$.

If $|\nabla z_l(\boldsymbol{r}_\parallel)|\ll 1$ then Eq. (5.56) is linearized[26], yielding [14]

$$q_{\mathrm{int.}}^{-2}\nabla^2 z_l(\boldsymbol{r}_\parallel)=z_l(\boldsymbol{r}_\parallel)-\int K(\boldsymbol{r}_\parallel-\boldsymbol{r}'_\parallel)\,z_s(\boldsymbol{r}'_\parallel)\,\mathrm{d}\boldsymbol{r}'_\parallel\,, \quad (5.57)$$

where $q_{\mathrm{int.}}$ and $K(\boldsymbol{r}_\parallel)$ are defined by

$$q_{\mathrm{int.}}^{-2}=\frac{\gamma}{\int w(\boldsymbol{r}'_\parallel,d)\,\mathrm{d}\boldsymbol{r}'_\parallel}\,,\qquad K(\boldsymbol{r}_\parallel)=\frac{w(\boldsymbol{r}_\parallel,d)}{\int w(\boldsymbol{r}'_\parallel,d)\,\mathrm{d}\boldsymbol{r}'_\parallel}\,. \quad (5.58)$$

Equation (5.57) can be solved in Fourier space, with the result

$$\tilde{z}_l(\boldsymbol{q}_\parallel)=\frac{\tilde{K}(\boldsymbol{q}_\parallel)}{1+\boldsymbol{q}_\parallel^2/q_{\mathrm{int.}}^2}\,\tilde{z}_s(\boldsymbol{q}_\parallel)\,. \quad (5.59)$$

Hence the replication factor $\chi(\boldsymbol{q}_\parallel)$ of a liquid film on a solid substrate is

$$\chi(\boldsymbol{q}_\parallel)=\frac{\tilde{K}(\boldsymbol{q}_\parallel)}{1+\boldsymbol{q}_\parallel^2/q_{\mathrm{int.}}^2}\,. \quad (5.60)$$

In the case of van der Waals interactions between the film and the substrate, i.e. $w(\boldsymbol{r}_\parallel,z)=-A_{\mathrm{eff}}\pi^{-2}/(r_\parallel^2+z^2)^3$, $q_{\mathrm{int.}}$ is simply the van der Waals cutoff $q_{\mathrm{int.}}=q_{\mathrm{vdW}}(d)=a/d^2$ with $a=\sqrt{A_{\mathrm{eff}}/(2\pi\gamma)}$, as previously defined in Eq. (3.20). For $\tilde{K}(\boldsymbol{q}_\parallel)$ one obtains

$$\tilde{K}(\boldsymbol{q}_\parallel)=\frac{q_\parallel^2\,d^2}{2}\,K_2(q_\parallel\,d)\,, \quad (5.61)$$

where $K_2(x)$ is the modified Bessel function of the second kind of order 2 [4, 147]. Since film thicknesses of $d\sim 100$ Å are typical and $q_\parallel$ is on the order of $q_\parallel\sim 10^{-5}$–10^{-3} Å^{-1} for a typical scattering experiment, the Bessel function in Eq. (5.61) may be expressed by its leading term, and thus $\tilde{K}(\boldsymbol{q}_\parallel)\simeq 1$. Hence

$$\chi(\boldsymbol{q}_\parallel)=\frac{a^2}{a^2+\boldsymbol{q}_\parallel^2\,d^4} \quad (5.62)$$

is obtained as the roughness replication factor for thin liquid films. The approximation $\tilde{K}(\boldsymbol{q}_\parallel)\approx 1$ is called the *Deryagin* approximation [83, 84].

It will be shown in the next chapter how the correlation functions enter the scattering theory. In particular, the scattering from surfaces with capillary waves will be discussed in some more detail.

[26] This is also known as the "linear response theory" or approximation.

6. Off-Specular Scattering

It was shown in Chap. 2 that rough interfaces damp the specularly reflected intensity considerably. This missing intensity is diffusely scattered at exit angles $\alpha_f \neq \alpha_i$, i.e. in off-specular directions[1]. Corrections of the specular scattering which take into account a continuous profile rather than sharp interfaces were discussed in Sect. 2.3. However, the specular reflectivity was found to be sensitive only to the vertical density profile $\varrho(z)$ of a sample[2]. Lateral fluctuations give rise to diffuse scattering, or vice versa: The diffusely scattered x-ray intensity provides information about lateral inhomogeneities of surfaces and interfaces.

In the past, great progress in the field of diffuse x-ray scattering has been achieved, both theoretically and experimentally. The next section provides a short overview of the theoretical developments. We shall not go into all details, because recently the reviews of *Dietrich & Haase* [94] and *Holý, Pietsch & Baumbach* [167] have appeared in the literature. Since long-range correlations are present on soft-matter surfaces the theory is slightly different from that for the scattering from solid interfaces. Therefore, the case of soft-matter surfaces is discussed separately in Sect. 6.1.3. Section 6.2 contains particular examples where off-specular x-ray scattering has provided new information about surfaces and interfaces that cannot be obtained by other methods. The specular reflectivity of all of these examples was discussed in Chap. 3. Thus, the gain of information can be seen more clearly.

6.1 Theory

The electron density of a sample is described in three dimensions by the function[3] $\varrho(x, y, z)$. It is convenient to separate the vertical profile $\varrho(z) = \langle \varrho(x, y, z) \rangle_{(x,y)}$ from the lateral fluctuations $\delta\varrho(x, y, z)$ in the following manner:

$$\varrho(x, y, z) = \varrho(z) + \delta\varrho(x, y, z) . \tag{6.1}$$

[1] Diffuse scattering is also found at the specular condition $\alpha_f = \alpha_i$.

[2] We neglect second-order DWBA effects (see Sect. 2.3).

[3] The same notation as in Chaps. 2 and 5 is used, with the z axis normal to the surface and (x, y) being the lateral coordinates.

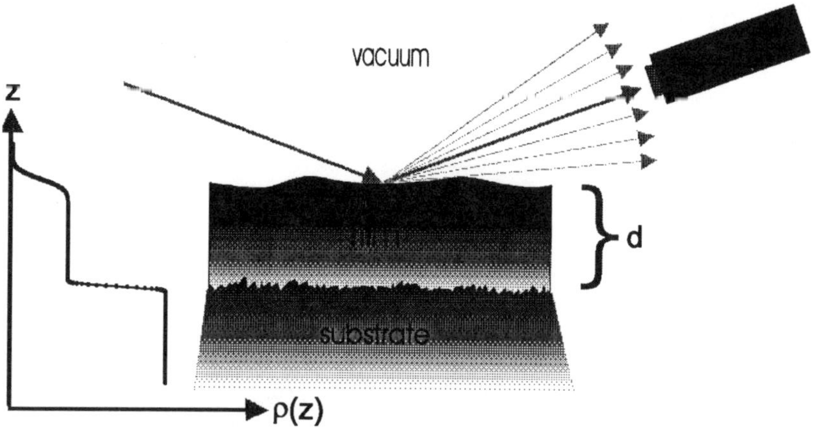

Fig. 6.1. Scattering of x-rays from a thin film of thickness d on top of a substrate: The radiation is specularly reflected (*thick arrow*), sampling the density profile (*left*) and scattered by lateral distortions (interface contours) in off-specular directions (*thin arrows*)

The vertical profile is sampled by the specular reflectivity and $\delta\varrho(x,y,z)$ accounts for the diffuse scattering [15, 94][4]. This situation is sketched in Fig. 6.1. Whereas the one-dimensional Helmholtz equation can be solved analytically for particular profiles $\varrho(z)$ (see Sect. 2.3), the full scattering problem including the diffuse scattering from the fluctuations $\delta\varrho(x,y,z)$ can only be solved by applying certain approximations.

Since the interaction of x-rays with matter is rather weak a simple kinematical treatment (see Sect. 4.1) has been very popular in the past. It was possible to explain the results of several diffuse x-ray scattering experiments with such a treatment (for examples see Refs. [194, 282, 314, 320, 321, 337, 338, 352]). In the next subsection the kinematical formulation of surface scattering is outlined for a single surface. The extension to multilayers is given afterwards within the framework of the more complex DWBA.

6.1.1 Kinematical Formulation

In the kinematical approximation, only single-scattering events are taken into account ("weak-scattering regime", see Sect. 4.1). Then the scattering function $S(\boldsymbol{q})$ is simply the modulus squared of the spatial Fourier transform of the scattering-length density $\varrho(\boldsymbol{r})$. This is equivalent to [335]

$$S(\boldsymbol{q}) = \iint \varrho(\boldsymbol{r})\varrho(\boldsymbol{r}')\exp\left\{\mathrm{i}\,\boldsymbol{q}(\boldsymbol{r}'-\boldsymbol{r})\right\}\mathrm{d}\boldsymbol{r}\,\mathrm{d}\boldsymbol{r}' , \tag{6.2}$$

[4] It is recapitulated here that the refractive index $n(x,y,z)$ is simply proportional to the electron density $\varrho(x,y,z)$. All further calculations are also valid for the case of neutron scattering; in this case the fuction $\varrho(x,y,z)$ represents the scattering-length density of the nuclei.

where the vectors $\boldsymbol{r} = (x, y, z)$ and $\boldsymbol{r}' = (x', y', z')$ are independent spatial coordinates, and $\boldsymbol{q} = (q_x, q_y, q_z)$ is the scattering vector as introduced in Sect. 3.1. The cross section, and hence the scattered intensity, is proportional to $S(\boldsymbol{q})$ (see Sect. 6.1.2).

The scattering function $S(\boldsymbol{q})$ of a single interface described by its contour $z(x, y)$ will be calculated explicitly. The density may be written as

$$\varrho(\boldsymbol{r}) = \Delta\varrho\, H[z - z(x, y)] \,, \tag{6.3}$$

where the step function[5] $H(z) = 0$ for $z \leq 0$, and $H(z) = 1$ for $z > 0$. The density contrast between the bulk values of the two materials next to the interface is $\Delta\varrho$. Inserting Eq. (6.3) into Eq. (6.2), and assuming that $z(x, y) - z(x', y')$ is a Gaussian random variable depending only on the distances $X = x - x'$ and $Y = y - y'$, we obtain[6]

$$\begin{aligned} S(\boldsymbol{q}) = \frac{(\Delta\varrho)^2}{q_z^2} \exp(-q_z^2\sigma^2) \iint \exp\left\{q_z^2\, C(X, Y)\right\} \\ \times \exp\left\{-\mathrm{i}(q_x X + q_y Y)\right\} \mathrm{d}X \mathrm{d}Y \,. \end{aligned} \tag{6.4}$$

Here σ is the rms roughness of the interface and $C(X, Y)$ is the height–height correlation function as described in the preceding chapter. Since $C(X, Y) \to 0$ as $X, Y \to \infty$, the integral in Eq. (6.4) contains a delta-function component. This point is more subtle for soft-matter surfaces and will be re-discussed in Sect. 6.1.3. For the moment the delta-function part is separated from the rest by writing the formula in the following way:

$$S(\boldsymbol{q}) = S_{\mathrm{spec.}}(\boldsymbol{q}) + S_{\mathrm{diff.}}(\boldsymbol{q}) \,, \tag{6.5}$$

where

$$S_{\mathrm{spec.}}(\boldsymbol{q}) = \frac{(\Delta\varrho)^2}{q_z^2} \exp(-q_z^2\sigma^2)\, \delta(\boldsymbol{q}_{\|}) \tag{6.6}$$

and

$$\begin{aligned} S_{\mathrm{diff.}}(\boldsymbol{q}) = \frac{(\Delta\varrho)^2}{q_z^2} \exp(-q_z^2\sigma^2) \\ \times \int \left[\exp\left\{q_z^2\, C(\boldsymbol{R})\right\} - 1\right] \exp(-\mathrm{i}\,\boldsymbol{q}_{\|} \cdot \boldsymbol{R})\, \mathrm{d}\boldsymbol{R} \,, \end{aligned} \tag{6.7}$$

are the explicit expressions for the specular and diffuse parts, respectively, with the lateral vectors $\boldsymbol{q}_{\|} = (q_x, q_y)$ and $\boldsymbol{R} = (X, Y)$.

The delta function in Eq. (6.6) restricts the scattering to the specular condition $\boldsymbol{q}_{\|} = (0, 0)$ and will not be discussed here (see Chap. 2 and Sect. 4.1).

From Eq. (6.7) some basic conclusions can be drawn. First, note that $S_{\mathrm{diff.}}(\boldsymbol{q})$ is proportional to $\Delta\varrho$. This means that the diffuse scattering is expected to be most pronounced for systems with a large electron density contrast. Hence almost all polymer/polymer interfaces, which have essentially no

---

[5] Also known as the Heaviside function.

[6] For details of this calculation see *Sinha et al.* [335] and Eq. (5.4).

electron density contrast, can be investigated neither by x-ray reflectivity nor by x-ray diffuse scattering. These interfaces are invisible in x-ray scattering unless special techniques are applied (contrast variation, etc.).

The second conclusion from Eq. (6.7) is that in general the diffuse scattering is *not* proportional to the PSD, i.e. the Fourier transform of $C(\boldsymbol{R})$. However, if the argument of the exponential $q_z^2 C(\boldsymbol{R})$ is small[7] then an expansion yields $S_{\text{diff.}}(\boldsymbol{q}) \sim \tilde{C}(\boldsymbol{q}_{\|})$. Only with this assumption, which is always true for light scattering but hardly ever for x-rays, is the diffuse intensity directly proportional to the PSD of an interface. It is interesting to note that detailed analysis of the scattering from liquid surfaces has confirmed the exponentials in Eqs. (6.4) and (6.7) [313]. This will be shown in Sect. 6.1.3.

As in the kinematical approximation all multiple-scattering effects are neglected, the derived formulas are not valid for $\alpha_i \approx \alpha_c$ or $\alpha_f \approx \alpha_c$. In the next subsection this shortcoming will be overcome.

6.1.2 Distorted-Wave Born Approximation

A common but already rather complicated treatment of the diffuse scattering has been performed in the past by applying the distorted-wave Born approximation (DWBA) to surface x-ray scattering problems. The DWBA is a combination of a dynamical and a kinematical treatment which exactly takes into account refraction at smooth interfaces, while the scattering at lateral inhomogeneities is treated kinematically, i.e. without inclusion of multiple-scattering effects. The scattering process is assumed to be a (small) perturbation of an ideal system. Then the cross section is calculated in first-order perturbation theory. Strictly speaking, instead of the "DWBA" the approximation should be called the "first-order DWBA".

The DWBA was first applied to grazing-angle x-ray-scattering problems by *Vineyard* [386] in 1982, and in the years 1983–1984 by *Dietrich & Wagner* [89, 90]. In 1988 *Sinha et al.* [335] for the first time calculated the cross section for diffuse x-ray scattering from a single surface within the DWBA[8]. This work was also fundamental with respect to the description of surfaces by the self-affine fractal model (see Sect. 5.3.1).

Later, in 1992, this result was confirmed by *Pynn* [290] and extended to scattering from a single layer. *De Boer* [71, 72] has shown that the inclusion of second order-effects in the theory avoids the principal problem of the first-order DWBA, namely that the total scattered intensity is not conserved (see also Sect. 2.3). In 1993 and 1994 *Holý & Baumbach* [164, 165] used the DWBA for layer systems, including the effect of vertical correlations between the interfaces. Such correlations are of decisive importance for many systems because in most thin-film preparation techniques (e.g. sputtering, evaporation, molecular beam epitaxy, LB film dipping) an imperfection in one layer

[7] More precisely, this means $q_z \sigma \ll 1$.

[8] A similar calculation but based on electrodynamic methods was given by *Andreev, Michette & Renwick* [15] in the same year.

is transferred to the layers above [263, 264, 282]. Recently *Dietrich & Haase* [94] and *Holý, Pietsch & Baumbach* [167] calculated the scattered intensity for various systems and experimental conditions. The interested reader will find the details of the calculations in these works. Appendix A.2 contains a brief introduction to the DWBA perturbation theory. Here, for sake of clarity, only the results will be discussed.

The expressions of the DWBA for the diffuse-scattering cross section are very similar to those obtained in the kinematical approximation (see Eq. 6.7). The main difference is that the transmission functions (see Eq. 2.9)

$$t_\mathrm{i}(\alpha_\mathrm{i}) = \frac{2\,k_{\mathrm{i},z}}{k_{\mathrm{i},z} + k_{\mathrm{t},z;\mathrm{i}}} \qquad \text{and} \qquad t_\mathrm{f}(\alpha_\mathrm{f}) = \frac{2\,k_{\mathrm{f},z}}{k_{\mathrm{f},z} + k_{\mathrm{t},z;\mathrm{f}}} \tag{6.8}$$

for the incident and outgoing radiation appear, where $k_{\mathrm{i},z} = k\sin\alpha_\mathrm{i}$, $k_{\mathrm{f},z} = k\sin\alpha_\mathrm{f}$, $k_{\mathrm{t},z;\mathrm{i}} = nk\sin\alpha_{\mathrm{t,i}} = k(n^2 - \cos^2\alpha_\mathrm{i})^{1/2}$, and $k_{\mathrm{t},z;\mathrm{f}} = nk\sin\alpha_{\mathrm{t,f}} = k(n^2 - \cos^2\alpha_\mathrm{f})^{1/2}$. Furthermore, the perpendicular wavevector transfer in the vacuum $q_z = k_{\mathrm{f},z} + k_{\mathrm{i},z}$ has to be replaced by that in the medium, $q_{z,\mathrm{t}} = k_{\mathrm{t},z;\mathrm{f}} + k_{\mathrm{t},z;\mathrm{i}}$. This accounts for refraction. The result for the diffusely scattered x-ray intensity $I_{\mathrm{diff.}}$ from a single surface, as obtained by *Sinha et al.* [335], is given by

$$I_{\mathrm{diff.}} \sim \left|t_\mathrm{i}(\alpha_\mathrm{i})\right|^2 S(\boldsymbol{q}) \left|t_\mathrm{f}(\alpha_\mathrm{f})\right|^2 , \tag{6.9}$$

with the scattering function

$$S(\boldsymbol{q}) = \frac{(\Delta\varrho)^2}{|q_{z,\mathrm{t}}|^2} \exp\left\{ - \left[(q_{z,\mathrm{t}})^2 + (q_{z,\mathrm{t}}^*)^2\right]\sigma^2/2\right\} \times \iint \left[\exp\{|q_{z,\mathrm{t}}|^2\, C(\boldsymbol{R})\} - 1\right] \exp(-\mathrm{i}\,\boldsymbol{q}_\parallel \boldsymbol{R})\, \mathrm{d}\boldsymbol{R} \,. \tag{6.10}$$

A distinct feature of the diffuse scattering is that maxima are observed whenever $\alpha_\mathrm{i} \approx \alpha_\mathrm{c}$ or $\alpha_\mathrm{f} \approx \alpha_\mathrm{c}$ (α_c is the critical angle). These maxima are caused by the transmission functions $t_\mathrm{i}(\alpha_\mathrm{i})$ and $t_\mathrm{f}(\alpha_\mathrm{f})$ in Eq. (6.9) and are called the *Yoneda* peaks [407][9]. It should also be noted that Eqs. (6.9) and (6.10) reduce to the kinematical result (see the previous subsection) if $q_{z,\mathrm{t}} \to q_z$ and $t_{\mathrm{i,f}} \to 1$, i.e. for large incidence and exit angles $(\alpha_\mathrm{i}, \alpha_\mathrm{f} > 3\,\alpha_\mathrm{c})$[10].

As an example Fig. 6.2 shows the diffuse scattering from a soft-matter thin film. A transverse scan along q_x is depicted and different characteristic regions are marked by numbers in parentheses. Region (1) indicates the background level that has to be subtracted before comparing the data with a scattering theory. In the regions (2) dynamical effects dominate the scattering. The pronounced maxima are the Yoneda peaks as explained above. Kinematical

[9] It was also proposed that the transmission functions for a rough interface (see Eqs. 2.35 and 2.41) should be used in Eq. (6.9) instead of $t_{\mathrm{i,f}}$ given by Eq. (6.8) [189, 391, 392]. A theoretical discussion of this point may be found in Ref. [73].

[10] This is not trivial, since the DWBA should, in principle, only be valid for small wavevector transfers!

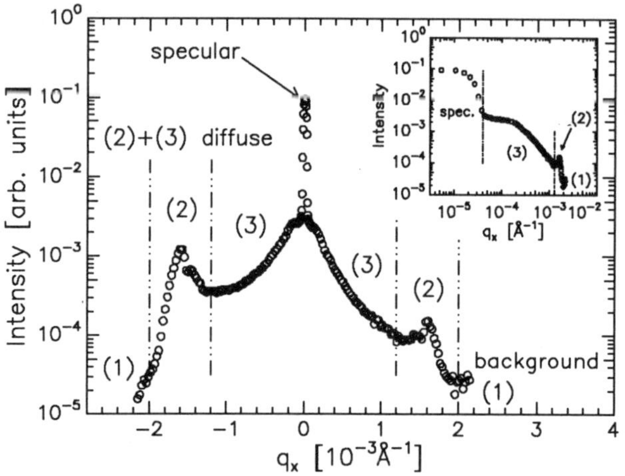

Fig. 6.2. Off-specular scattering (*open circles*) from a PEP film of thickness $d = 200$ Å on a Si/SiO_2 substrate. The inset shows a log–log plot of the right-hand side of this scan, emphasizing the kinematical region (3). The other regions are (1) the background level, and (2) the regions where dynamical scattering dominates

scattering is the main contribution in the regions (3). Here the transmission functions in Eq. (6.9) may be approximated by unity. In the center, at $q_x = 0$, the sharp resolution-limited specular peak is visible. The inset in the figure depicts a log–log plot of the right-hand part of the curve. In such a plot the region of kinematical scattering is highlighted. As we shall see in the next subsection this kind of plot is appropriate whenever capillary waves are expected.

The extension of the result given by Eqs. (6.9) and (6.10) to layer systems as calculated by *Holý & Baumbach* [165], yields for the diffusely scattered intensity (for the notation see Chap. 2)

$$I_{\text{diff.}}(\boldsymbol{q}) = \frac{\mathcal{G}k^2}{8\pi^2} \sum_{j,k=1}^{N} \left(n_j^2 - n_{j+1}^2\right)\left(n_k^2 - n_{k+1}^2\right)^* \times \sum_{m,n=0}^{3} \widetilde{G}_j^m \widetilde{G}_k^{n*} \exp\left\{-\tfrac{1}{2}\left[\left(q_{z,j}^m \sigma_j\right)^2 + \left(q_{z,k}^{n*}\sigma_k\right)^2\right]\right\} S_{jk}^{mn}(\boldsymbol{q}) \,, \tag{6.11}$$

where $\mathcal{G}$ is the illuminated area, and the scattering function $S_{jk}^{mn}(\boldsymbol{q})$ is given by

$$S_{jk}^{mn}(\boldsymbol{q}) = \frac{1}{q_{z,j}^m q_{z,k}^{n*}} \int \left(\exp\left\{q_{z,j}^m q_{z,k}^{n*}\, C_{jk}(\boldsymbol{R})\right\} - 1 \right) \exp(-\mathrm{i}\, \boldsymbol{q}_\parallel \cdot \boldsymbol{R})\, \mathrm{d}\boldsymbol{R} \,. \tag{6.12}$$

The perpendicular wavevector transfers $q_{z,j}^m$ and $q_{z,k}^n$ inside the layers j and k appear here. These wavevector transfers and the quantities $\widetilde{G}_j^m = G_j^m \exp(-\mathrm{i}\, q_{z,j}^m \mu_j)$ are given explicitly in Appendix A.2.

The lateral shape of the interfaces enters into the formulas via the height–height correlation functions $C_j(\boldsymbol{R}) = C_{jj}(\boldsymbol{R})$ and hence into the scattered intensity. Vertical roughness correlations between interfaces are taken into account by the respective cross-correlation functions $C_{jk}(\boldsymbol{R})$ $(j \neq k)$. Finally, it should be mentioned that the integral given by Eq. (6.12) is difficult to calculate numerically. This makes the data analysis very time-consuming. However, certain approximations for this integral exist in the literature [94, 269, 338, 354, 355].

Equation (6.12) also shows that x-ray scattering from surfaces and interfaces may be better treated in real space via correlation functions than in reciprocal space with PSDs. The next subsection shows some consequences for soft-matter surfaces.

6.1.3 Soft-Matter Surfaces

The scattering from surfaces with capillary waves will be discussed in more detail here. For bulk liquids, Eq. (6.4) in Sect. 6.1.1 is used for the evaluation, where a separation into a specular and a diffuse component is *not* performed. Adapting Eq. (6.4) to experimental conditions, with wide-open slits in the out-of-plane y direction[11] and an inclusion of the q_x resolution δ_{q_x} via a real-space cutoff function $R(X) = \exp(-4\pi^2 X^2/L^2)$ with $L = 2\pi/\delta_{q_x}$ (for details see Sect. 3.1.1), yields the kinematical scattering function

$$S(q_x, q_z) = \frac{(\Delta\varrho)^2}{q_z^2} \exp(-q_z^2\sigma^2) \times \int_0^\infty \exp\left\{q_z^2\, C(X)\right\} \exp\left(- 4\pi^2 X^2/L^2\right) \cos(q_x X)\, \mathrm{d}X \,, \tag{6.13}$$

using the rms roughness σ and the one-dimensional height–height correlation function $C(X) = C(X, 0)$. Equation (6.13) is still a general and exact expression for $S(\boldsymbol{q})$. If the particular form $C(X) = (B/2)K_0(q_{\mathrm{l,c}}X)$ (see Eq. 5.23) were inserted, a separation between a delta-like purely specular part and a diffuse contribution would indeed be possible since $K_0(q_{\mathrm{l,c}}X) \to 0$ for $X \to \infty$. However, since the cutoff $q_{\mathrm{l,c}} = q_{\mathrm{g}} = \sqrt{g\,\varrho/\gamma} \sim 10^{-7}$–$10^{-8}$ Å^{-1} is very small for bulk liquids, and in practice $q_{\mathrm{g}} \ll \delta_{q_x}$ is always fulfilled, $K_0(q_{\mathrm{l,c}}X)$ may be well approximated by its leading term, i.e. $K_0(q_{\mathrm{l,c}}X) \approx -\ln(q_{\mathrm{l,c}}X/2) - \gamma_{\mathrm{E}}$ [4, 147], as discussed in Sect. 5.3.3. This approximation is valid throughout the experimentally accessible q range.

After inserting the logarithmic correlation function in Eq. (6.13) and multiplying by the respective transmission functions $|t_{\mathrm{i}}(\alpha_{\mathrm{i}})|^2 |t_{\mathrm{f}}(\alpha_{\mathrm{f}})|^2$ to account for dynamical scattering effects (see Eq. 6.9)[12], the following result was obtained by *Sanyal et al.* [313] for the observed intensity $I(q_x, q_z)$ recorded in a detector:

[11] This means an integration over q_y.

[12] This is not quite correct, since the resolution is not included for t_{i} and t_{f}.

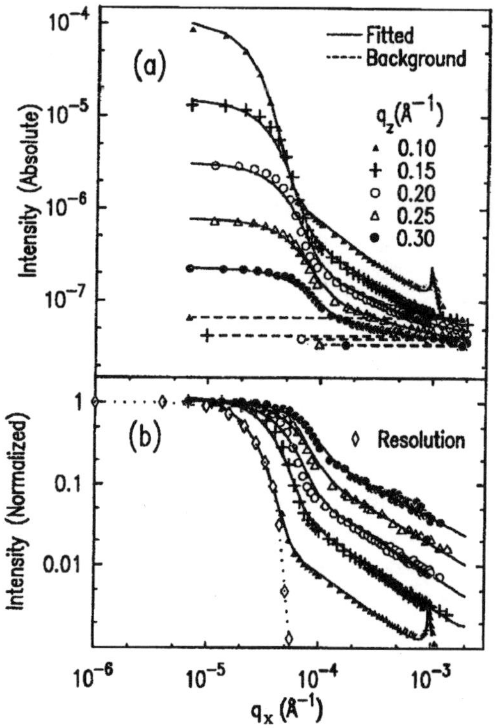

Fig. 6.3. **(a)** Log–log plot of the diffusely scattered intensity (*symbols*) vs. q_x from a free ethanol surface at several values of q_z. The *lines* are fits to the data (*dashed lines*: backgrounds) with a model assuming free capillary waves without a cutoff (logarithmic correlation function). The peak at $q_x \approx 10^{-3}$ Å^{-1} is the Yoneda peak due to dynamical scattering. **(b)** Log–log plot of the transverse diffuse scattering normalized to unity at $q_x = 0$ for the q_z values of (a). The *lines* are fits to the data and the *dashed curve* indicates a scan of the main beam profile. The characteristic q_z dependence of the power-law exponent can be seen. This q_z dependence is a fingerprint of capillary waves on liquid surfaces (figure taken from *Sanyal et al.* [313])

$$I(q_x, q_z) = I_0 \left(\frac{q_c}{2\, q_z} \right)^4 \frac{\sin\alpha_i + \sin\alpha_f}{2\sqrt{\pi}\,\sin\alpha_i} \exp(-q_z^2\sigma_{\text{eff}}^2)\,\Gamma\Big[(1-\eta)/2\Big]$$
$$\times\, {}_1F_1\Big[(1-\eta)/2\,;1/2\,;q_x^2L^2/(4\pi^2)\Big]\, \Big|t_i(\alpha_i)\Big|^2 \Big|t_f(\alpha_f)\Big|^2 , \qquad (6.14)$$

where I_0 is the incident beam intensity, $q_c = 2k\sin\alpha_c$ is the wavevector corresponding to the critical angle α_c, $\Gamma(x)$ is the Gamma function, ${}_1F_1(x;y;z)$ is the Kummer function [4, 105, 147], and L may be considered as the effective coherence length along the surface (see Sect. 3.1.1). The effective surface roughness σ_{eff} is given by Eqs. (3.10) and (3.11), and η was set to

$$\eta(q_z) = \frac{k_B T}{2\pi\gamma} q_z^2 = \frac{1}{2} B q_z^2 , \qquad (6.15)$$

with the same parameter B that was used in Sects. 3.1.1 and 5.3.3.

In the specular case, $\alpha_i = \alpha_f$ ($q_x = 0$), Eq. (6.14) may be simplified to[13] $I(0, q_z) = I_0 R_F \exp(-q_z^2\sigma_{\text{eff}}^2)$, where R_F is the Fresnel reflectivity as introduced in Sect. 2.1 (see Eq. 2.12). Hence the damping of the specu-

[13] Note that ${}_1F_1(x;y;0) = 1$. The factor $(1/\sqrt{\pi})\Gamma[(1-\eta)/2]$ is essentially unity.

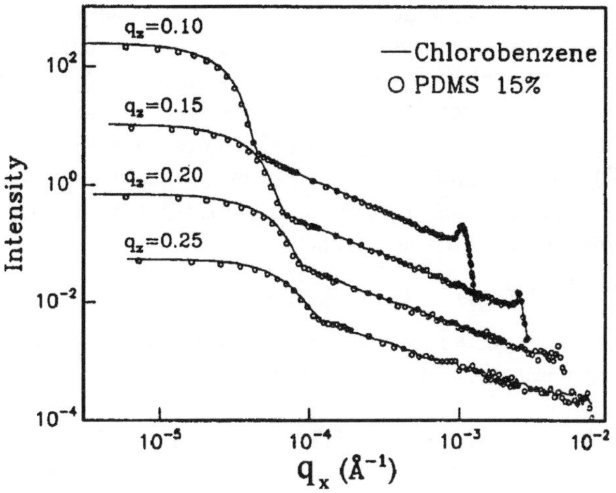

Fig. 6.4. Diffuse scattering from pure chlorobenzene (*lines*) and a 15% PDMS/chlorobenzene solution (*open circles*). Log–log plots of q_x scans at different q_z values are shown. The units of q_z are Å^{-1}. The feature at $q_x \approx 10^{-3}\,\text{Å}^{-1}$ is the Yoneda peak due to dynamical scattering (figure taken from *Zhao et al.* [412])

larly reflected intensity with a *Beckmann–Spizzichino* factor is regained (see Sects. 2.3 and 3.1.1 and Fig. 3.5).

Equation (6.14) is also the expression for the diffusely scattered intensity. For $q_x \gg \delta_{q_x} = 2\pi/L$ and fixed q_z, an expansion of the Kummer function yields [4, 105, 147]

$$I_{\text{diff.}}(q_x, q_z) \sim q_x^{\eta(q_z)-1} \,. \tag{6.16}$$

Hence, capillary wave fluctuations on soft-matter surfaces yield particular power laws for the diffuse scattering. The main feature is that the power $\eta - 1$ depends on the vertical momentum transfer q_z by virtue of Eq. (6.15) [145, 396]. The first experimental proof of this prediction was given by *Sanyal et al.* [313] in 1991. Figure 6.3 depicts diffuse-scattering data from a free ethanol surface (beamline X22B, NSLS Brookhaven, $\lambda = 1.52\,\text{Å}$)[14]. It can be seen that the Kummer function of Eq. (6.14) provides the smooth transition between the Gaussian-like "specular" part and the power-law tails with q_z-dependent exponents according to Eqs. (6.15) and (6.16).

Another example is shown in Fig. 6.4. Here log–log plots of the diffusely scattered intensity versus q_x are presented for pure chlorobenzene and a 15% PDMS/chlorobenzene solution. Again the power-law behavior is clearly seen, leading one to suspect that free capillary waves are present on the surface even for polymer solutions in the semidilute region [412].

Now extensions to the case of thin films will be made. In Sect. 5.3.3 a simple replacement of the gravitational cutoff $q_{\text{l,c}} = q_g$ by the much larger van der Waals cutoff $q_{\text{l,c}} = q_{\text{vdW}}(d) = a/d^2$, with the length $a \sim 5$–$10\,\text{Å}$, was discussed (see Eq. 3.20). Since in general $q_{\text{vdW}}(d)$ is also larger than

[14] Figure 3.5 in Sect. 3.1.1 shows the "specular" and longitudinal diffuse scattering from the same system together with calculations based on Eq. (6.14).

the resolution, a specular part is contained in Eq. (6.13) and a separation according to Eqs. (6.5)–(6.7) into specular and diffuse scattering is regained. Thus, for thin films the replacement of the Bessel function $K_0(q_{l,c}X)$ by the logarithm (see above) is not valid for all q_x and the integral in Eq. (6.7) or Eq. (6.12) has to be evaluated numerically with the correlation function $C(X) = (B/2)K_0(q_{l,c}X)$ for the fluid interface. However, the logarithmic expansion of $K_0(q_{l,c}X)$ is still valid for $q_{l,c}X \ll 1$, and a similar calculation to that given above for bulk liquid surfaces yields

$$I_{\text{diff.}}(q_x, q_z) \sim q_x^{\eta(q_z)-1} \quad \text{for} \quad q_x \gg q_{l,c} = q_{\text{vdW}}(d) \; . \tag{6.17}$$

Hence, the same power law is expected for large q_x. In the region $\delta_{q_x} < q_x < q_{l,c}$ a more complicated expression is found that is dominated by higher-order terms of $K_0(q_{l,c}X)$.

One has to evaluate the integrals in Eqs. (6.7) or (6.12) numerically for most of the other correlation functions discussed in Chap. 5. Therefore we will not go into any more details here, except for the case of micro-islands on PEP surfaces, where the correlation function is given empirically by Eq. (5.44) (see Sect. 5.3.3). The scattering for large q_x will be further discussed analytically. From Eq. (5.44) the following expansion of the correlation function may be deduced [4, 147]:

$$C(X) \approx \frac{B}{2}\left\{ -\ln\left(\frac{q_{l,c}X}{2}\right) - q_0 X - \gamma_E + \frac{\pi\, q_0}{2\, q_{l,c}} \right\} + O(q_{l,c}^2 X^2) \; , \tag{6.18}$$

which is valid for $q_{l,c}X \ll 1$. Inserting Eq. (6.18) into Eq. (6.13) and noticing that $L \gg 1/(\eta\, q_0)$ is a rather good approximation, the following $S(\boldsymbol{q})$ is obtained:

$$S(q_x, q_z) \approx \frac{(\Delta\varrho)^2}{q_z^2} \exp(-q_z^2\sigma_{\text{eff}}^2) \left(\frac{2}{q_{l,c}}\right)^{\eta} \left(q_x^2 + q_{l,c}^2 + \eta^2 q_0^2\right)^{(\eta-1)/2}$$
$$\times\, \Gamma(1-\eta)\, \cos\left[(1-\eta)\arctan\left(\frac{q_x}{\sqrt{q_{l,c}^2 + \eta^2 q_0^2}}\right)\right] \; , \tag{6.19}$$

where $\eta = Bq_z^2/2$ and $q_0 = 1/d$ are defined as before in Sect. 5.3.3. The combination $\sqrt{q_{l,c}^2 + \eta^2 q_0^2}$ has been inserted instead of ηq_0 to obtain the correct limit in the case $\eta \to 0$[15]. From Eq. (6.19), the following asymptotic behaviour may be obtained:

$$S(q_x, q_z) \sim q_x^{\eta(q_z)-1} \quad \text{for large } d \; , \tag{6.20}$$
$$S(q_x, q_z) \sim q_x^{\eta(q_z)-2} \quad \text{for small } d \; , \tag{6.21}$$

which is valid for $q_x \gg q_{\text{eff}}(q_z) = \sqrt{q_{l,c}^2 + \eta^2 q_0^2}$. Thus, the particular q_z-dependent capillary wave exponent changes for thin films from $\eta - 1$ to $\eta - 2$,

[15] For $\eta \to 0$ (i.e. $q_z \to 0$) the scattering function $S(\boldsymbol{q})$ must be equal to the PSD.

and also the onset $q_{\mathrm{eff}}(q_z)$ of the power-law region is now predicted to be q_z-dependent.

Qualitatively, the change of the capillary wave exponent is simply the consequence of the fact that the scattering is proportional to the Fourier transform of $\exp\{q_z^2 C(R)\}$. Here, for a surface possessing capillary waves plus islands, $C(R) = C_{\mathrm{cap}}(R) + C_{\mathrm{is}}(R)$ (see Eq. 5.45 in Sect. 5.3.3), thus leading to

$$S(q_x, q_z) = \frac{(\Delta\varrho)^2}{q_z^2} \exp(-q_z^2 \sigma_{\mathrm{eff}}^2)\, S_{\mathrm{cap}}(q_x, q_z) * S_{\mathrm{is}}(q_x, q_z) \,, \tag{6.22}$$

where $*$ denotes a convolution in q_x, and $S_{\mathrm{cap}}(q_x, q_z) = \int \exp\{q_z^2 C_{\mathrm{cap}}(X) + \mathrm{i}\, q_x X\} \mathrm{d}X$ and $S_{\mathrm{is}}(q_x, q_z) = \int \exp\{q_z^2 C_{\mathrm{is}}(X) + \mathrm{i}\, q_x X\} \mathrm{d}X$ are (q_x, q_z)-dependent scattering functions due to capillary waves and islands, respectively. Hence, one expects the island asymptotic power-law behavior decorated with the capillary wave q_z-dependent exponent as described by Eqs. (6.20) and (6.21).

At this point the discussion of the scattering theory will be stopped. Some further details will be provided in the following sections in connection with experiments.

6.2 Experiments

In contrast to reflectivity experiments, off-specular-scattering studies from soft-matter films are rather rare. This may be due to the scattering theory, which is much more difficult than the simple Fresnel theory and hence complicates the data analysis considerably. In Sect. 6.2.3 a particular example is discussed, proving this statement. There, a detailed quantitative analysis of the scattering from LB films is given.

In the two subsections before, 6.2.1 and 6.2.2, diffuse scattering from polymer films and liquid films is discussed. The specular reflectivity of these systems has already been presented in Sect. 3.2. Hence, the additional information that can be extracted from the off-specular data can be seen. The last example, Sect. 6.2.4, deals with roughness propagation in soft-matter thin films. Here a special technique was used, where surface gratings serve as substrates to separate unambiguously the conformal part of the roughness from the intrinsic capillary waves. Experimental details are omitted in the description since the setups already described in Sect. 3.1 were used.

Like the specular reflectivity, the diffuse-scattering cross section is proportional to the scattering-length density contrast at an interface (see Sect. 6.1). Hence, as before, polymer/polymer interfaces do not contribute to the scattering in the case of x-rays. Off-specular neutron scattering, where a high contrast between two polymers may be easily achieved by deuteration, is often affected by the low intensity of neutron sources. This limits the number of off-specular neutron-scattering studies (e.g. Refs. [111, 117, 213, 362]).

6.2.1 Polymer Films

An interesting class of polymers, which has been successfully investigated in the past by off-specular scattering, consits of freestanding films. As already mentioned in Sect. 3.2.1, *Daillant & Bélorgey* [68, 69] presented an x-ray-scattering study on ultrathin soap films where attention was focused on diffuse scattering, both theoretically and experimentally. In 1995, *Shindler et al.* [334] investigated correlations in the thermal fluctuations of freestanding smectic-A films[16]. This first quantitative study shows how powerful off-specular x-ray scattering is to obtain structural information from systems which are hardly accessible by other techniques, although they are well understood theoretically [168]. Recently, in 1997, *Mol et al.* [81, 246, 247, 248, 249] extended these investigations to the determination of displacement–displacement correlations of freely suspended films, from mesoscopic down to molecular in-plane distances. While for long-wavelength undulations of the smectic film the top and bottom undulate conformally, a crossover to independent fluctuations for shorter wavelengths was detected in accordance with theories based on the Landau–de Gennes free-energy density expression [74, 77, 381]. This example also demonstrates another strength of x-ray scattering: Mesoscopic and atomic dimensions are accessible in the same experiment.

Examples of thin PS and PEP films on silicon will be explicitly discussed in this subsection. The specular reflectivities and the properties of these systems have already been described in Sect. 3.2.1 (see Figs. 3.9–3.13). In fact, the analysis of the specular reflectivity already contains information that was obtained from the off-specular data because the explanation of the roughness curves $\sigma_{\mathrm{PS}}(d)$ (see Eq. 3.21 and Fig. 3.11) and $\sigma_{\mathrm{PEP}}(d)$ (see Eq. 3.25 and Fig. 3.13) was based on an integration over the respective PSDs (see Sect. 5.2).

From AFM images sampling a region of $20 \times 20\,\mu\mathrm{m}^2$ an rms roughness of $\sigma_{\mathrm{AFM}} \approx 3\,\mathrm{\AA}$ for all PS samples was obtained. This is considerably smaller than the value obtained by x-ray reflectivity. Moreover, with the AFM a logarithmic roughness increase with the film thickness as shown in Fig. 3.11 was not detectable, and Fourier transformations of the images did not yield PSDs with characteristic power laws. As was shown in Sect. 3.2.1 and will be shown below, scattering methods – in particular x-ray reflectivity and diffuse scattering – yield more reliable results for fluctuating soft-matter interfaces. One reason may be the sampled area, which is even larger than $20 \times 20\,\mu\mathrm{m}^2$ for x-ray scattering using grazing angles. Since long-range correlations may be present on soft-matter surfaces the extent of the sampled area should have a drastic influence on the measured rms roughness (see Eq. 5.11). However, the values of σ_{AFM} were almost unaffected by the size of this area in the present study. Nevertheless the real-space information of AFM measurements

[16] For the physics and a review of works related to smectic films see e.g. the book by *Chandrasekhar* [57] and references therein.

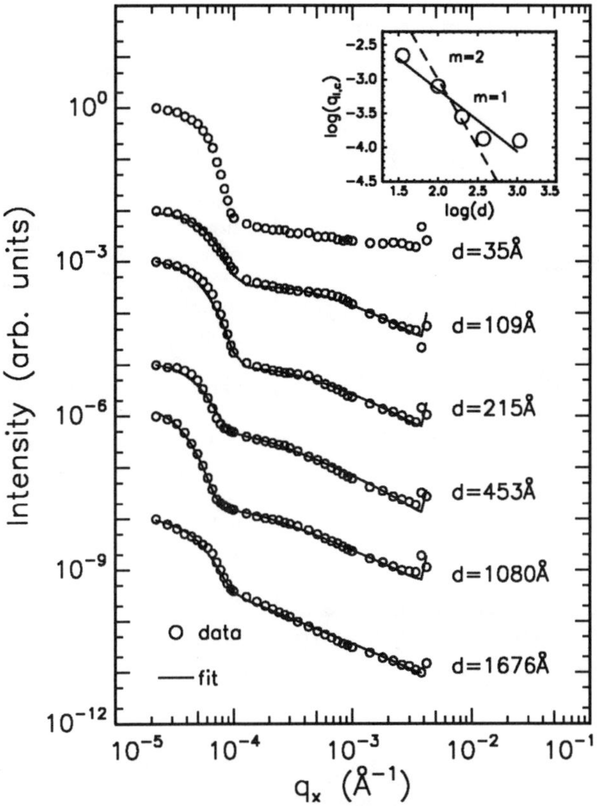

Fig. 6.5. Diffuse scattering from polystyrene films on silicon substrates with film thicknesses d = 35–1676 Å. *Open circles*: transverse scans along q_x for fixed $q_z = 0.2\,\text{Å}^{-1}$. *Lines*: fits with a lower cutoff $q_{l,c}$ as parameter (not for the d = 35 Å film). The inset depicts $\log(q_{l,c})$ vs. $\log(d)$. *Open circles*: $q_{l,c}$ obtained from the diffuse-scattering data. *Solid line*: linear fit, yielding a slope of $m = 1$ corresponding to $q_{l,c} = a/d$. *Dashed line*: linear relationship corresponding to $m = 2$ according to a pure van der Waals cutoff $q_{l,c} = a/d^2$ [378]

is important because it gives a quick overview over the expected length scales. The importance of AFM will become clearer when the measurements of the PEP films are discussed (see Fig. 6.8).

Figures 6.5 and 6.6 depict diffuse-scattering data from the series of PS films. The corresponding reflectivities have already been shown in Fig. 3.10. Log–log plots of transverse scans along q_x at fixed $q_z = 0.2\,\text{Å}^{-1}$ are presented for films within the thickness range d = 35–1676 Å.

Qualitatively, the log–log plots can be explained in the following way: The region of very small q_x contains the specular peak with a Gaussian-like shape (the resolution). The thinnest film in Fig. 6.5 (d = 35 Å) shows an almost constant level of diffuse scattering that is quite similar to that of the bare silicon substrate. No calculation was done for this film.

For the thicker films characteristic power laws are observed, quite similar to those expected for capillary waves (see Sect. 6.1.3, Eqs. 6.16 and 6.17) [28]. However, the situation is more subtle than for free capillary waves on bulk liquid surfaces. Since the lower cutoff $q_{l,c}$ is expected to be much larger for a thin film than the gravitational cutoff (see Sect. 5.3.3), the full Bessel-

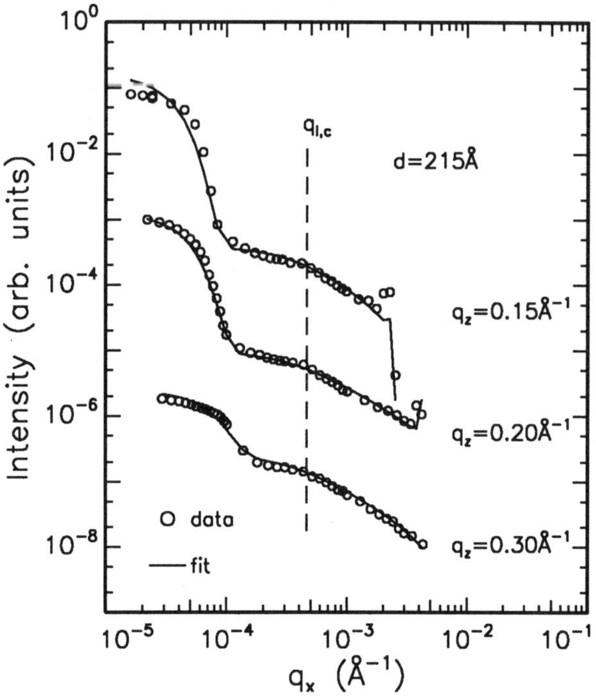

Fig. 6.6. Transverse scans along q_x at different q_z values for a polystyrene film of thickness $d = 215$ Å on silicon. *Open circles*: data. *Lines*: fits. *Vertical broken line*: location of the lower cutoff $q_{l,c}$. The location of the cutoff is *not* a function of q_z. The region of very small q_x is dominated by the Gaussian-shaped specular peak, which broadens with increasing q_z. The q_z dependence of the power-law exponent in the region $q_x > q_{l,c}$ is weak [378]

function expression $C(X) = (B/2)K_0(q_{l,c}X)$ given by Eq. (5.23) has to be used as the correlation function in the calculations, and not the simple logarithmic expansion. The lines in Figs. 6.5 and 6.6 are calculations, where the lower cutoff was a free fit parameter. It can be seen that the cutoff moves to smaller q_x with increasing film thickness and that the q_x location of the cutoff shows no q_z dependence (see vertical broken line in Fig. 6.6). For the thickest film with $d = 1676$ Å, this cutoff is already hidden under the specular peak, similarly to the case of bulk liquid surfaces. Therefore, the data can be explained by a simple replacement of the lower gravitational cutoff $q_{l,c} = q_g$ by a larger thickness-dependent cutoff as described in Sect. 5.3.3.

However, quantitatively there are significant discrepancies between the experimentally obtained $q_{l,c}$ values and those given by a van der Waals cutoff $q_{l,c} = q_{vdW}(d) = a/d^2$. This can be seen in the inset of Fig. 6.5, where a log–log plot of $q_{l,c}$ versus the film thickness d is shown. A linear (solid line) fit reveals a slope of $m = 1$ instead of the predicted $m = 2$ (dashed line) due to $q_{vdW}(d) \propto d^{-2}$.

Another discrepancy is found for the parameter $\eta = (B/2)q_z^2$ (see Sect. 6.1.3) which controls the slope of the power-law region (see Eqs. 6.15–6.17) if capillary waves are present. The fits reveal (see Fig. 6.6) that η is

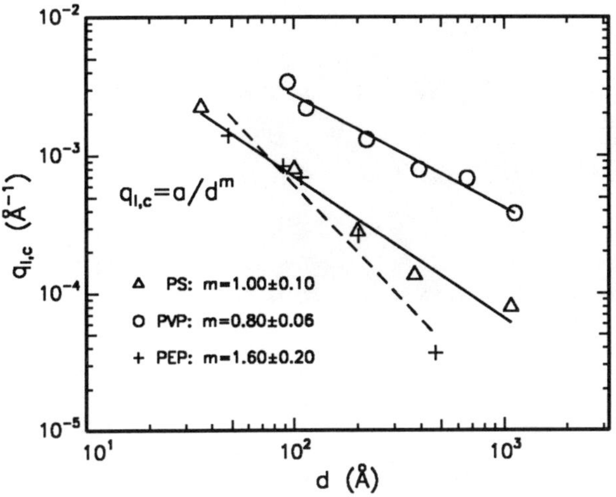

Fig. 6.7. Log–log plot of the lower cutoffs $q_{l,c}$ as a function of the film thickness d for different polymers on silicon substrates. If a relationship $q_{l,c} = a/d^m$ is assumed then the power $m = 0.8$ results for PVP, $m = 1.0$ for PS, and $m = 1.6$ for PEP, agreeing with a decreasing strength of the substrate/adsorbate interactions [378]

almost zero. This would correspond to a very small B value and hence to a much larger surface tension γ of PS than the value known from the literature.

Explanations for this enhanced surface tension may be given by the curvature corrections or pair potential corrections as discussed in Sect. 5.3.3 [95, 257, 258, 259] (see Eqs. 5.33 and 5.34). However, the roughness curve $\sigma_{PS}(d)$ was well explained by the assumption $q_{l,c} = q_{vdW}(d)$ but with a lowered surface tension (see Fig. 3.11). Therefore, more observed features can be explained within the *Fredrickson* model [128], as outlined in Sect. 5.3.3 (see Eqs. 5.38–5.41). This model explains (i) the logarithmic increase of the roughness curve with the film thickness where the bulk surface tension can be used (see Sect. 3.2.1), (ii) the correct power-law dependence $q_{l,c} \propto d^{-1}$ of the cutoff (see Eq. 5.39 and Fig. 6.7), but *not* (iii) the small η values which are almost independent of q_z. However, the first two points suggest that the film is in a highly viscous state, almost corresponding to a brush-type structure which is assumed in the *Fredrickson* model. Capillary waves seem to play a minor role on PS thin-film surfaces.

This explanation is supported by Fig. 6.7, where the location of the cutoffs $q_{l,c}$ are plotted on a log–log scale as a function of the film thickness d for the three different polymers PS, PVP, and PEP on Si substrates. If a power law $q_{l,c} = a/d^m$ is assumed, then one obtains $m = 0.80 \pm 0.06$ for PVP, in excellent agreement with the *Fredrickson* model (see Eq. 5.41), which predicts $m \approx 0.75$, since the interaction of silicon with PVP is even stronger than that with PS. On the other hand, PEP does not stick on silicon, a pure van der Waals behavior corresponding to $m = 2$ is expected, and $m = 1.6 \pm 0.2$ is found. These measurements will now be discussed with the focus on another point since PEP also dewets an oxide-covered silicon surface [376, 413].

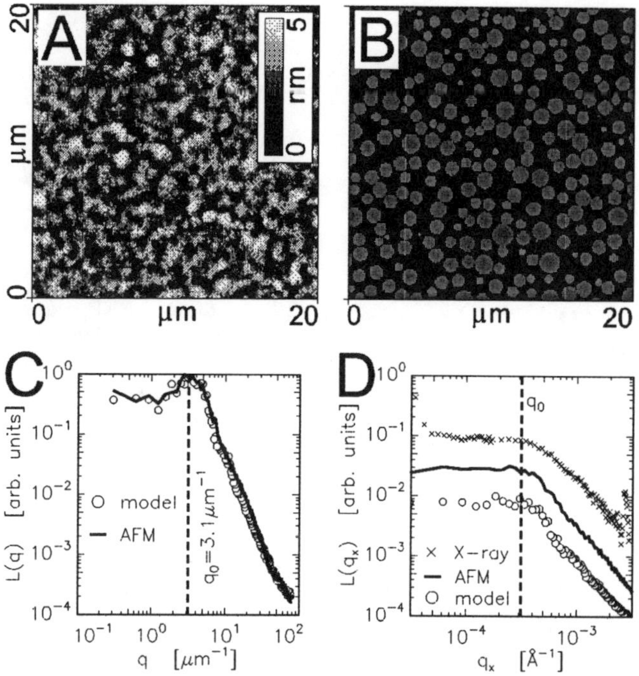

Fig. 6.8. **(A)** AFM image of a PEP surface in the initial stage of dewetting, for a film of thickness $d = 107$ Å. **(B)** Sketch of a surface possessing a random arrangement of circular islands. **(C)** *Line*: PSD (here denoted by $L(q)$ instead of $\tilde{C}(q_{\|})$) as obtained from a 2D Fourier transform of the AFM image and a subsequent angular average. *Open circles*: PSD obained by Fourier-transforming the arrangement shown in image (B). **(D)** *Line*: PSD $L(q_x)$ [$= \tilde{C}(q_x)$] averaged over q_y as calculated from the AFM image. *Open circles*: PSD $L(q_x)$ averaged over q_y as calculated with the arrangement shown in (B). *Crosses*: x-ray data obtained with a coarse resolution in the q_y direction [376]

A large body of work, both theoretical and experimental, has been done concerning the lateral mesoscopic structures that form during dewetting. Mostly, polymer films were the subject of the investigations (see e.g. Refs. [39, 175, 237, 253, 254, 255, 291, 292, 295, 305, 405, 413]). However, little is known about the microscopic structure of the surfaces on top of the islands of dewetted films. PEP films on silicon surfaces that were in the initial capillary instability stages of dewetting, where the sample surface is just roughened on a mesoscopic scale (islands) but a complete dewetting down to the substrate has not yet occurred, were investigated. A discussion of the specular reflectivities and a description of the preparation of the PEP samples was given in Sect. 3.2.1. Since the glass transition temperature of PEP is $T_G \approx -62°C$ the films are expected to be in the liquid state[17].

The investigations were done in two steps: (i) Images of the PEP surfaces were taken with a Digital Instruments 3000 AFM in the noncontact tapping mode. These images reveal the characteristic island structures of dewetted films [413]. (ii) X-ray diffuse-scattering experiments were performed (HASYLAB, beamline E2; the geometry is sketched in the inset of Fig. 6.10). The x-ray scattering signal is sensitive to the microscopic part of the roughness, too, i.e. in this case to possible capillary wave fluctuations.

Figure 6.8A shows the AFM data for the $d = 107$ Å thick PEP film. Quite similar structures were found on all film surfaces. The island size increases with increasing film thickness, as observed in detail by *Zhao et al.* [413] and in accordance with theoretical predictions [49]. The line in Fig. 6.8C depicts the PSD (here denoted by $L(q)$ instaed of $\tilde{C}(q_{\|})$) obtained by Fourier-transforming the AFM image. Since no orientation is preferred, the angular average is plotted. A maximum at $q_0 = 3.1\,\mu m^{-1}$ can be seen, which is caused by the regular mesoscopic structure of the islands (see Sect 5.3.3). The open circles in Fig. 6.8C correspond to a numerically calculated PSD for the simulated island distribution in Fig. 6.8B where circular islands, all of the same height and without microscopic roughness, were assumed. The result of the simulation and the AFM image agree well, both in real and in reciprocal space. Figure 6.8D (line and open circles) shows the respective PSDs but now averaged over q_y instead of the isotropic angular dimension. Owing to the coarse out-of-plane resolution, this quantity corresponds essentially to what is measured in an x-ray experiment (crosses). It is worth noting that this average transforms the maximum at q_0 into a cutoff-like "kink".

The discussion above shows that the AFM data can be fully explained by the mesoscopic island structure but further details are invisible. The main reason for this is that details on an atomic scale of fluctuating interfaces are hard to observe with an AFM, where a tip is moved over the surface. Here scattering techniques have significant advantages.

Figures 6.9 and 6.10 depict the data (circles) and fits (lines) of transverse scans taken along the direction q_x (see inset of Fig. 6.10). The data shown in Fig. 6.9 were measured at $q_z = 0.2\,\text{Å}^{-1}$. In the region $q_x < 5 \times 10^{-5}\,\text{Å}^{-1}$ the specular peak was modeled by a Gaussian. The diffuse scattering, i.e. the intensity measured for $q_x > 5 \times 10^{-5}\,\text{Å}^{-1}$, was fitted with Eqs. (6.19) and (6.22) given at the end of Sect. 6.1.3. Two regions can be identified: (i) A flat region close to the specular peak. The size of this region decreases with increasing film thickness. (ii) A power-law region where the diffusely scattered intensity is proportional to $q_x^{\eta-2}$. Figure 6.10 shows that the exponent is a function of q_z and the fits (lines) reveal the q_z-dependent exponents typical of capillary waves. However, because of the islands, the additional correlation

[17] The glass transition temperature T_G may, however, be strongly dependent on the film thickness (see the discussion in Sect. 3.2.1).

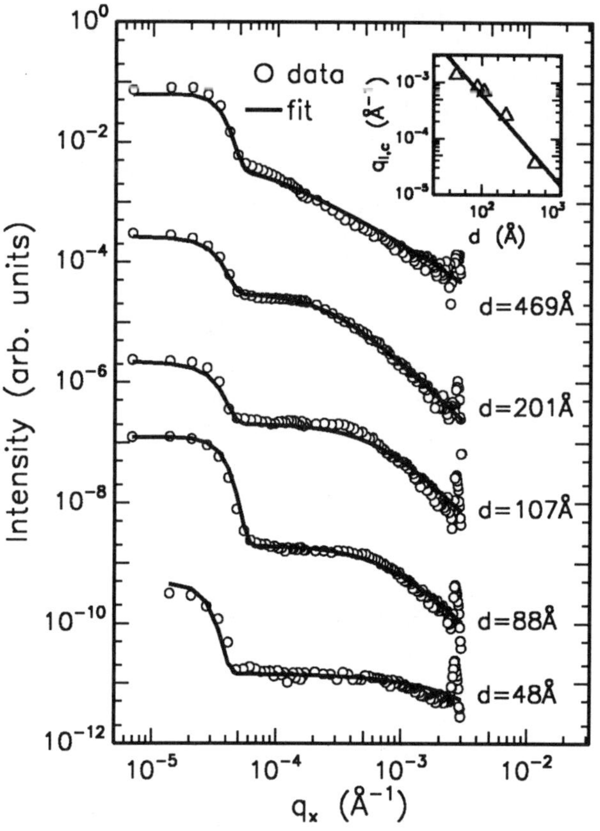

Fig. 6.9. Diffuse scattering from PEP films on silicon substrates with film thicknesses d = 48–469 Å. *Open circles*: transverse scans for fixed $q_z = 0.2\,\text{Å}^{-1}$. *Lines*: fits. The inset depicts a log–log plot of the lower capillary wave cutoff $q_{l,c}$ vs. d. *Open triangles*: lower cutoffs $q_{l,c}$ obtained from the diffuse-scattering fits. *Line* linear fit yielding $q_{l,c} = a/d^{1.6\pm0.2}$. The Yoneda peak at $q_x \approx 3 \times 10^{-3}\,\text{Å}^{-1}$ is caused by dynamical scattering effects and was excluded from the data analysis [376]

function $C_{\text{is}}(R)$ (see Eq. 5.45 in Sect. 5.3.3) yields a $q_x^{\eta-2}$ behavior instead of the usual pure capillary wave form $q_x^{\eta-1}$ [313] (see dashed lines in Fig. 6.10). The inclined vertical dashed line emphasizes that the cutoff q_{eff} is a function of q_z, too, as anticipated by the calculations of the scattering in Sect. 6.1.3. Only for the thickest film is the whole diffuse scattering essentially determined by a power-law region with $q_x^{\eta-1}$ (see Eqs. 6.20 and 6.21). Thus, for this thickness, free capillary waves which propagate on the liquid PEP surface are already found.

Here it should be emphasized again that the fits shown in Fig. 6.10 confirm the q_z dependence of the exponent in the power-law region. This particular form is a fingerprint of a logarithmically diverging correlation function, i.e. it proves unambiguously the presence of capillary waves on the liquid PEP surfaces [313, 335].

The inset of Fig. 6.9 depicts a log–log plot of $q_{l,c}$ versus the film thickness. The straight line, with slope $-1.6 \approx -2$, suggests that a van der Waals cutoff $q_{l,c} = a/d^2$ with $a \approx 7\,\text{Å}$ can be identified from the data. From the nominal

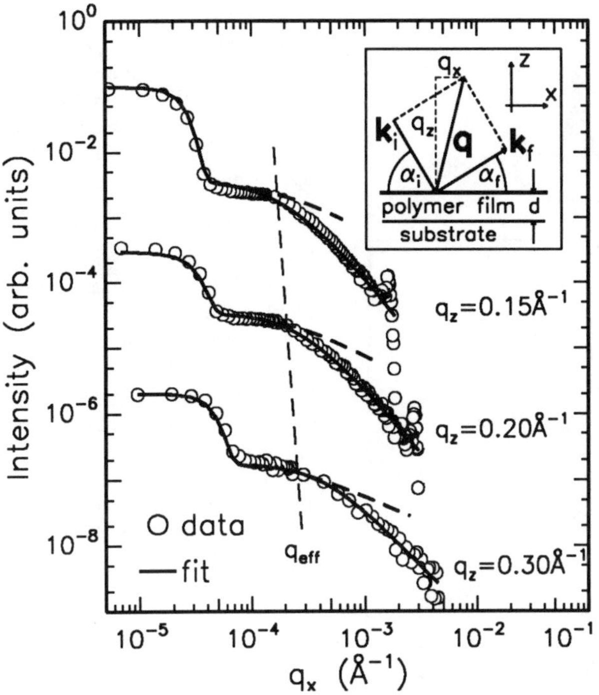

Fig. 6.10. Transverse scans along q_x at q_z values 0.15 Å^{-1}, 0.20 Å^{-1}, and 0.30 Å^{-1} for a PEP film of thickness d = 201 Å on silicon. *Open circles*: data. *Lines*: fits with the island–capillary wave model. *Inclined vertical dashed line*: location of the q_z-dependent cutoff $q_{eff}(q_z)$. The other *dashed lines* show calculations for capillary wave surfaces in the absence of islands. The inset depicts the scattering geometry [376]

parameters for PEP, $A_{eff} \sim 10^{-18}$–10^{-19} J for the effective Hamaker constant and $\gamma = 30\,\text{dyn/cm}^2$ for the surface tension [413], a value of a = 7–20 Å may be calculated, again showing a good agreement between data and theory.

As already discussed in Sect. 3.2.1, the theory of Brochard-Wyart et al. [49] for resonant capillary modes which are present at the pre-rupture stage of an unstable film in the initial stage of dewetting yields a power-law behavior for the resonant wavenumber q_0. The data analysis suggests $q_0 \propto d^{-n}$ with $n \approx 1$. Since the error bars are quite large the theoretical value of $n = 3/2$ is still within the accuracy of the experiment.

From the discussion of the AFM and off-specular x-ray scattering data of PEP two major conclusions can be drawn: (i) The PEP surfaces were in the initial stages of dewetting. (ii) The propagation of capillary waves on the liquid PEP surfaces is hindered only by the van der Waals background potential and is superimposed on the island substructures. The two effects were separated in a most direct and unambiguous manner.

This last example reveals that combined real-space and reciprocal-space investigations can provide a complete picture of a sample, including fluctuations of the surface and substrate/adsorbate interactions. Another example of a combined STM/x-ray study is one of C_{60} islands on VSe_2. A complete surface/interface investigation of this system in real and reciprocal space was

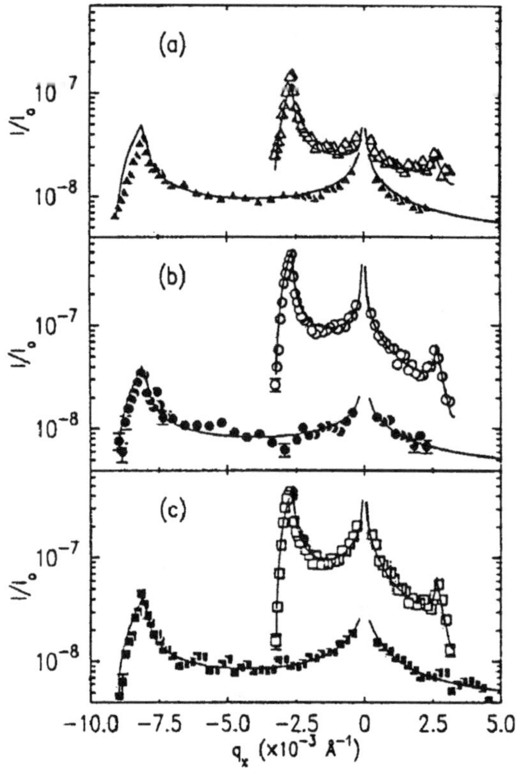

Fig. 6.11. Off-specular scattering (transverse scans) of three liquid cyclohexane films on silicon. The corresponding reflectivities are shown in Fig. 3.20 (see Sect. 3.2.2). The *open symbols* were measured at $q_z = 0.1558\,Å^{-1}$ and the *solid symbols* at $q_z = 0.2597\,Å^{-1}$. For the sake of clarity the latter data have been multiplied by 0.5. The film thicknesses are **(a)** $d = 22\,Å$, **(b)** $d = 44\,Å$, and **(c)** $d = 128\,Å$. Fits to the data are indicated by *lines*. The specular peak at $q_x = 0$ is omitted and the peaks on the left and right are the Yoneda peaks due to dynamical scattering (figure taken from *Tidswell et al.* [365])

recently performed by *Schwedhelm et al.* [328]. The next subsection presents experiments on liquid films consisting of small molecules (i.e. "real liquids") – systems where scattering methods are almost the only tool to obtain structural information at atomic dimensions.

6.2.2 Liquid Thin Films

As already mentioned in Sect. 3.2.2, in 1991 *Tidswell et al.* [365] presented the first diffuse-scattering study of thin liquid films on solid substrates. The aim of this work was to show (i) that roughness replication according to the theory of *Andelmann, Joanny & Robbins* [14, 298] takes place in thin liquid cyclohexane films, and (ii) that the diffuse scattering can be explained by the capillary wave model with van der Waals cutoff (for details see Sect. 5.3.3).

It was found that very thin films with $d < 60\,Å$ replicate the roughness, while the surfaces of thicker films were unaffected by the roughness of the substrate. This was visualized in Fig. 3.20 in Sect. 3.2.2. Figure 6.11 depicts transverse scans for two different q_z values, for the cyclohexane films whose reflectivities were shown in Fig. 3.20. The lines are calculations with

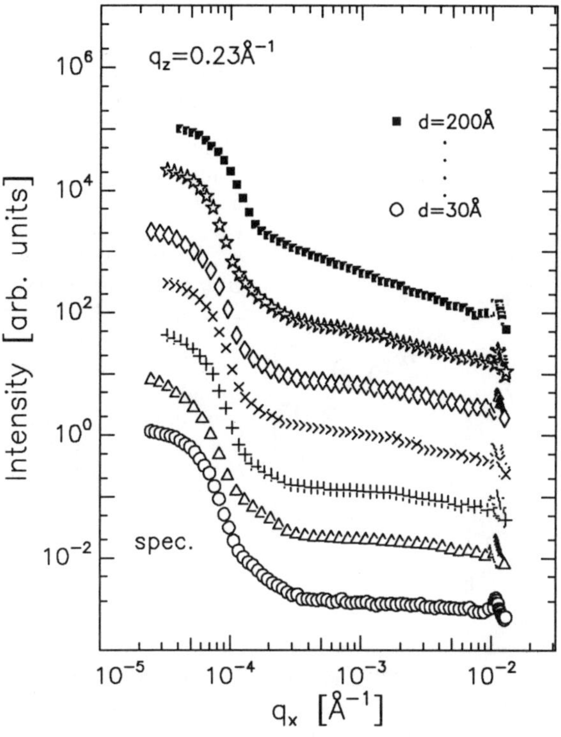

Fig. 6.12. Transverse diffuse-scattering scans from thin liquid hexane films adsorbed on silicon substrates. The film thicknesses are d = 30–200 Å. At very small q_x the Gaussian-shaped specular peak can be seen, and the feature at $q_x \approx 10^{-2}$ Å^{-1} is the Yoneda peak. The scattering of the thickest film exhibits a power law corresponding to capillary waves with a van der Waals cutoff. All curves are shifted for clarity. Scattering from a bare wafer would be two orders of magnitude below the scattering level of the thinnest film [96]

a simplified scattering theory and coincide well with the data. However, the interpretation of the scattering data was somewhat difficult since the scattering theory used did not contain the characteristic power laws discussed in Sect. 6.1.3. Thus, the presence of capillary waves cannot be unambiguously proven by the fits[18]. Also, the separation of the replicated from the intrinsic roughness is rather difficult since the scattering from the two parts has essentially the same $q_{||}$ dependence (see Eqs. 5.34 and 5.62).

Later, *Seeck et al.* [330] presented an x-ray diffuse-scattering study of a thin CCl_4 film on silicon. Again, capillary waves were not proven by power laws. However, evidence for capillary waves was nevertheless found by a detailed data refinement.

The diffuse scattering of a series of liquid hexane films is presented in Fig. 6.12 [96]. The corresponding reflectivities were discussed in Sect. 3.2.2. No power laws could be identified for the thinner films. Only the thickest film, with d = 200 Å, shows this signature of capillary waves. The data are shown separately in Fig. 6.13. Two transverse scans obtained at different perpendicular wavevector transfers q_z are depicted. The diffuse scattering

[18] Unfortunately, the scattering from the silicon substrate was also assumed to obey a power law.

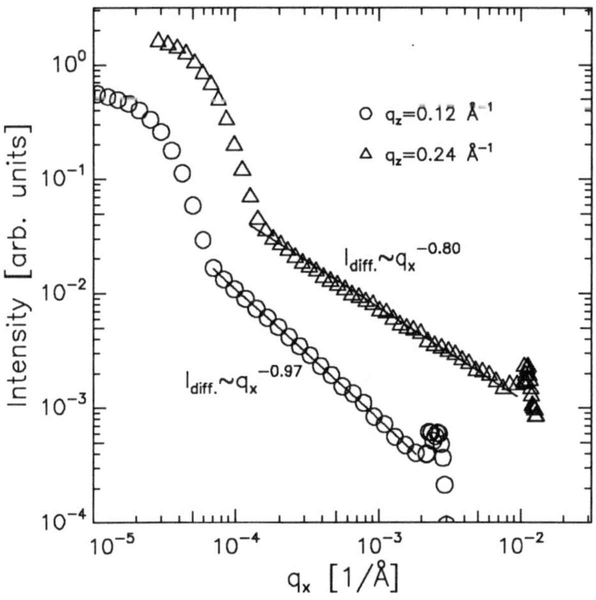

Fig. 6.13. Power laws of the diffuse tails for a liquid hexane film of thickness $d = 200$ Å. Log–log plots of two transverse scans along q_x for fixed $q_z = 0.12$ Å^{-1} (*open circles*) and $q_z = 0.24$ Å^{-1} (*open triangles*) are shown. The *lines* are linear fits of the power-law region. The Gaussian-like specular part can be seen on the *left* and the Yoneda peak due to enhanced transmission on the *right* [96]

can be explained with a simple power law $I_{\text{diff.}} \sim q_x^{\eta-1}$ (lines) where the exponent η shows the correct quadratic q_z dependence (see Eqs. 6.15–6.17). Hence, free capillary waves seem to be present at the surface of this film.

Since neither power laws nor clear cutoffs (see e.g. Fig. 6.5) were observed, a simple application of the capillary wave models presented in Sect. 5.3.3 is impossible. One explanation is a very strong substrate/film interaction suppressing capillary waves. The density profiles which were obtained from the reflectivity data (see Sect. 3.2.2 and Fig. 3.22) suggest that the near-substrate region of the hexane films is strongly affected by the substrate/film interactions, hence supporting this argument. However, the extent of the low-density region was limited to 50 Å. It should be noted, too, that in contrast with the cyclohexane study of *Tidswell et al.* [365] (see above), no conformal roughness was observed even for the thinnest hexane films.

For films of thicknesses $d > 50$ Å capillary waves may indeed be present. This was suggested by the logarithmic increase of the surface roughness with the film thickness which was obtained from the analysis of the reflectivity experiments (see Fig. 3.23). However, the signatures in the diffuse scattering do not unambiguously prove the presence of capillary waves on the thinner liquid hexane films.

These polymer and liquid-film examples have shown how capillary waves may be detected at surfaces. The logarithmic correlation function and the particular film thickness dependence of the cutoff are signatures of this type

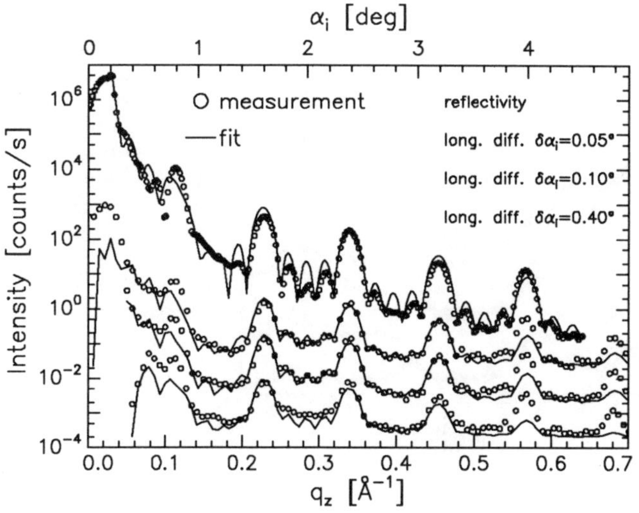

Fig. 6.14. Total reflectivity (specular and diffuse) and three longitudinal diffuse scans for offsets $\delta\alpha_i = 0.05°, 0.10°, 0.40°$ for a nine-layer CdA LB film. *Open circles*: measurements. *Lines*: fits. For clarity, all curves are shifted by one order of magnitude on the intensity scale [264]

of surface wave. In the next subsection the diffuse scattering from another class of soft-matter films is discussed, namely LB films.

6.2.3 Langmuir–Blodgett Films

In contrast to reflectivity experiments only a few diffuse-scattering studies of LB film structures can be found in the literature [27, 140, 263, 264, 356, 357, 358]. The reflectivity experiments of Sect. 3.2.4 revealed vertically well-ordered LB films. However, the reflectivity alone could not detect the high degree of conformal roughness that was found for CdA multilayers (see Fig. 3.28). In this subsection the respective diffuse scattering measurements will be discussed, with particular emphasis on roughness correlations.

As the first example, an LB film consisting of nine CdA layers on top of an oxide-covered silicon wafer is discussed [264]. All measurements shown in Figs. 6.14–6.16 were made using a rotating-anode laboratory source ($\lambda = 1.54$ Å). The total reflectivity (Fig. 6.14) was measured to a maximum wavevector transfer of $q_z = 1.4\,Å^{-1}$. As already noticed in Sect. 3.2.4 for other CdA films, the reflected intensity is purely diffuse for large q_z values (here, for $q_z > 0.7\,Å^{-1}$). The distance $\Delta q_z = 0.113\,Å^{-1}$ between the Bragg peaks in Fig. 6.14 corresponds to a Cd–Cd spacing of $2\pi/\Delta q_z = 55$ Å, i.e. to the thickness of one bilayer, confirming the Y-structure of the CdA LB film (see inset of Fig. 3.29) [140, 264]. The rapid oscillations ($\Delta q_z = 0.025\,Å^{-1}$) correspond to the total thickness of $d = 251$ Å of the whole layer stack.

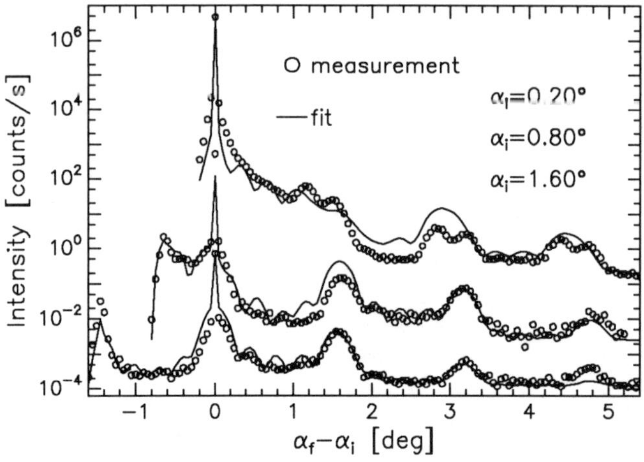

Fig. 6.15. Detector scans at three different angles of incidence $\alpha_{\rm i} = 0.20°$, $0.80°$, $1.60°$ for the nine-layer CdA sample. *Open symbols*: measurement. *Solid lines*: fits. For clarity all curves are shifted on the intensity scale [264]

More detailed information about the interfaces was obtained by measurements and fits of the diffuse scattering. In Fig. 6.14 three longitudinal diffuse scans (see Sect. 3.1 and Figs. 3.3 and 3.4) with offset angles $\delta\alpha_{\rm i} = 0.05°, 0.10°, 0.40°$ are also shown.

Figures 6.15 and 6.16 show three detector scans with fixed angles of incidence $\alpha_{\rm i} = 0.20°, 0.80°, 1.60°$ and eight rocking curves (transverse scans) within the region $0.050\,\text{Å}^{-1} \leq q_z \leq 0.229\,\text{Å}^{-1}$. All scans taken together yield a representative picture of the diffusely scattered intensity in reciprocal space. With a regular slab model[19] and the scattering formulas given by Eqs. (6.11) and (6.12) in Sect. 6.1.2 the fits (solid lines) in Figs. 6.14–6.16 were obtained. Self-affine height–height correlation functions $C_j(R)$ were assumed for all interfaces (see Eq. 5.12 in Sect. 5.3.1). The roughness correlations were taken into account by cross-correlation functions $C_{jk}(R)$ according to Eq. (5.55) in Sect. 5.4, with a vertical correlation length $\xi_\perp$ [244]. Hence all interfaces are assumed to have the same lateral correlation length ξ and the same Hurst parameter h. Note that the correlation function of a self-affine fractal interface given by Eq. (5.12) has to be considered here as a parametrization of the "real" but unknown $C_j(R)$. On the length scales that in-plane diffuse x-ray scattering probes, the interfaces appear self-affine. But it is unrealistic to assume self-affine interfaces for smaller length scales (see also Ref. [264]).

The structure of the three longitudinal diffuse scans in Fig. 6.14 is well reproduced by the fits. Their pronounced modulations indicate strong vertical correlations between the interfaces of the different layers. This can be qualitatively understood by the following simple consideration. If identical functions $C_{jk}(R)$ are used in Eqs. (6.11) and (6.12) for the calculation of the

[19] Each CdA bilayer was split into two very thin Cd layers ($d_{\rm Cd} \approx 2.5$ Å) with high density and a long low-density hydrocarbon chain ($d_{\rm CH} \approx 50$ Å).

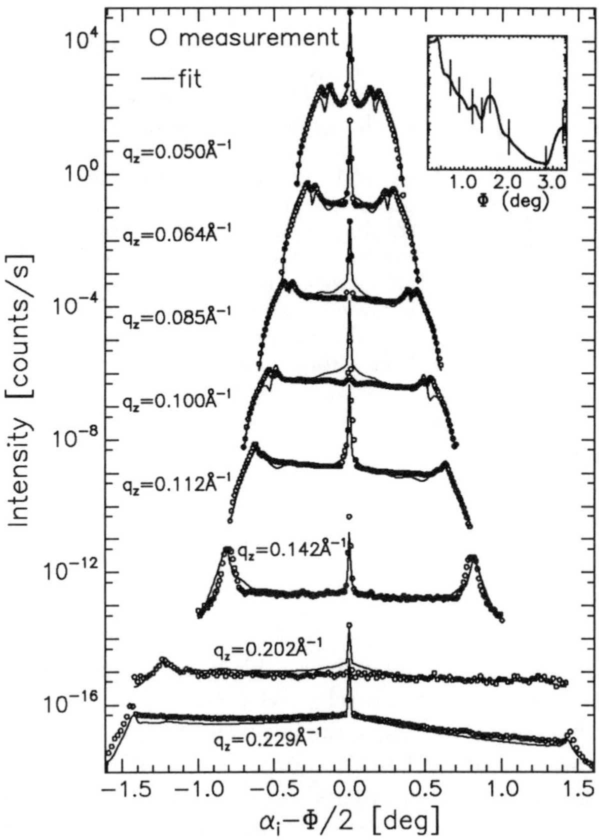

Fig. 6.16. Transverse scans at different q_z positions for the nine-layer CdA sample. $\Phi = \alpha_i + \alpha_f$ denotes the scattering angle, i.e. the angle between the position of the detector and the incoming beam. *Symbols*: measurements. *Solid lines* best fits. For clarity all curves are displaced by one order of magnitude on the intensity scale. In the middle a narrow specular component is visible. This component vanishes for $q_z > 0.7\,\text{Å}^{-1}$. The inset shows the low-angle region of the reflectivity and the positions at which the transverse scans are taken [264]

scattering and kinematical scattering is assumed, then the resultant expression for the diffuse scattering contains exactly the same phase factors as the reflectivity. Hence, all oscillations of the reflectivity are also found in diffuse scattering scans in the case of totally correlated interfaces [338].

A more quantitative description is given by the vertical correlation length $\xi_\perp$ which determines the degree of conformality in the model described by Eq. (5.55). A value of $\xi_\perp = 700\,\text{Å}$ was obtained from the refinements. This is almost three times larger than the total LB film thickness of $d = 251\,\text{Å}$ and is an indication that the vertical roughness correlations are rather perfect. Also, the modulation amplitudes and periods are nearly independent of the chosen offset angle $\delta\alpha_i$, proving that shorter wavelengths ($\approx 2000\,\text{Å}$) of the roughness spectrum are transferred in a nearly undamped fashion from the bottom to the top through the whole LB film [133].

The detector scans shown in Fig. 6.15 and the transverse scans of Fig. 6.16 are also well reproduced by the calculations. Hence the simple slab model is able to explain all diffuse-scattering data quantitatively. Almost perfect

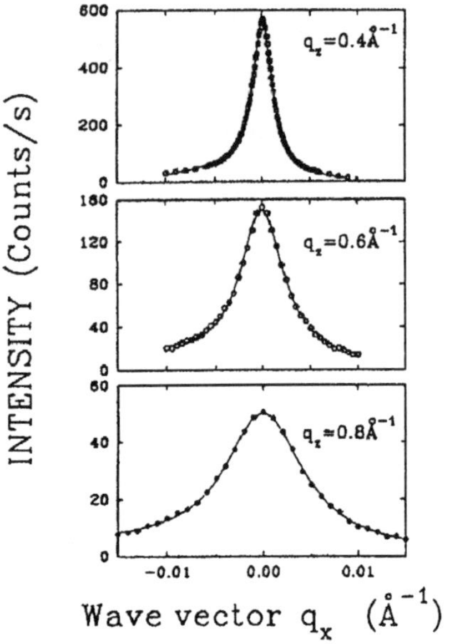

Fig. 6.17. Transverse scans through the one-dimensional Bragg peaks of the reflectivity profile shown in Fig. 3.29 (see Sect. 3.2.4). The data are shown by the *symbols* and the *lines* are fits of Lorentzian line shapes. An increase of the FWHM Δq_x with increasing perpendicular wavevector transfer q_z according to $\Delta q_x \propto q_z^2$ is found. This can be quantitatively explained by self-affine rough interfaces which are perfectly correlated and where no further lateral cutoff is introduced (figure taken from *Gibaud et al.* [140])

vertical roughness correlations are also indicated by the detector scans, since modulations corresponding to those of the reflectivity are visible even for larger exit angles. From the transverse scans, the in-plane correlation length $\xi \approx 300\,\text{Å}$ was obtained together with a Hurst parameter $h = 0.5$. The length ξ was found to correspond to the size of ordered regions as probed by GID measurements [263, 264].

This example shows that the effort required to perform a complete analysis of specular and diffuse scattering is rather large. However, the result is a characterization of the interfaces of organic multilayers by only a few parameters[20], which may be optimized by variation of the preparation conditions. Recently, *Basu & Sanyal* [27, 317] observed power laws in the diffuse scattering from LB films, indicating that long-range correlations quite similar to those of capillary waves may be found, too.

Another study of a nine-layer CdA LB film, published by *Gibaud et al.* [140], for which the reflectivity has already been discussed in Sect. 3.2.4 (see Fig. 3.29), yielded similar results. An analysis of the peak width Δq_x of transverse scans at the q_z positions of the Bragg peaks for large q_z values was performed. Figure 6.17 depicts three scans for $q_z = 0.4, 0.6, 0.8\,\text{Å}^{-1}$, together with Lorentzian fits.

The increase of Δq_x can be explained by assuming perfectly correlated self-affine rough interfaces without a cutoff. Thus, all scattering is essentially

[20] Rms roughnesses of the interfaces also were determined.

diffuse and a calculation using Eq. (6.4) in Sect. 6.1.1 yields the intensity $I(q_x, q_z)$ at a point (q_x, q_z) of reciprocal space [140, 335]:

$$I(q_x, q_z) = \int S(\boldsymbol{q})\, \mathrm{d}q_y = \frac{2\, A\pi\, |F(\boldsymbol{q})|^2}{q_x^2 + (A/2)^2\, q_z^4}\,, \tag{6.23}$$

where $F(\boldsymbol{q})$ is the structure factor of the Bragg peaks. The out-of-plane resolution was assumed to be completely relaxed, and as the height difference function[21] the pure self-affine form $g(\boldsymbol{R}) = A\, R^{2h}$ was used with a roughness exponent $h = 0.5$. Equation (6.23) shows that the intensity distribution for a transverse scan is a Lorentzian of width $\Delta q_x = A\, q_z^2$. This quadratic q_z dependence of Δq_x was observed for the CdA system, with $A = 0.016\,\text{Å}$ [140]. The value of A is relatively small compared to the length scale of the multilayer structure. This means that, on the one hand, the disorder within the system is not large. On the other hand, it is highly correlated and transferred from layer to layer. It is interesting that even for very rough interfaces no attenuation factor like a *Beckmann–Spizzichino* or *Névot–Croce* factor (see Sect. 2.3) in Eq. (6.23) appears. This is the reason why LB film Bragg peaks of highly correlated interfaces possessing very large correlation lengths can be observed up to very high perpendicular wavevector transfers (here up to $q_z \approx 2\,\text{Å}^{-1}$, see Fig. 3.29).

These two examples of soft-matter multilayers show that x-ray scattering is well suited to detect the high degree of roughness replication in those samples. For future applications of LB films as sensors it is of great importance to understand the mechanism of roughness replication and coupling of the individual layers within a multilayer structure, since rather perfect systems are required.

6.2.4 Roughness Propagation in Soft-Matter Films

Vertical roughness correlations in LB films were discussed in the previous section in terms of a real space description via correlation functions. It was shown that x-ray scattering is a powerful tool to obtain a quantitative description of the degree of conformality in such systems. However, for many other systems two major problems arise: (i) Often it is quite difficult to separate the conformal part from the intrinsic roughness since the scattering due to the latter may be much weaker than in the case of highly correlated LB multilayers. (ii) The full application of the scattering theory outlined in Sect. 6.1.2 (see Eqs. 6.11 and 6.12), even in the simple kinematical approximation, is rather time-consuming. Thus, a way is presented here to investigate vertical roughness propagation in soft-matter thin films in a more systematic and fundamental manner.

The start of the considerations is the replication factor $\chi_{jk}(\boldsymbol{q}_{\|})$ as defined by Eq. (5.50) in Sect. 5.4. The subscript "jk" will be omitted in the following

[21] Note that $g(\boldsymbol{R}) = 2\sigma^2 - 2C(\boldsymbol{R})$.

because a single film on top of a substrate will be treated. There are only a few specific formulations of replication factors given in the literature which have been calculated on the basis of microscopic models. *Spiller, Stearns & Krumrey* [346] derived the following $\chi(q_\parallel)$ from a particular growth model:

$$\chi(q_\parallel) = \exp(-q_\parallel^2\, \nu\, d)\ , \tag{6.24}$$

with a relaxation parameter ν and a layer thickness d, from which in the case of multilayers the number of correlated layers $N(q_\parallel) = 1/\{2\ln(1+q_\parallel^2 \nu d)\}$ may be deduced. The quantity $N(q_\parallel)$ gives the number of bilayers of thickness $2d$ that is required to damp a roughness component with wavevector $q_\parallel$ to $1/e$ of its original amplitude. Equation (6.24) has been tested by x-ray scattering experiments for several systems, e.g. W/C and Si/Ge multilayers [306, 310, 312, 345, 373].

As shown in Sect. 5.4 for thin liquid films, *Andelmann, Joanny & Robbins* [14, 298] calculated the replication factor $\chi(q_\parallel)$ in the linear-response approximation:

$$\chi(q_\parallel) = \frac{a^2}{a^2 + q_\parallel^2\, d^4}\ , \tag{6.25}$$

with the length $a = \sqrt{A_{\mathrm{eff}}/(2\pi\gamma)}$ as defined before by Eq. (3.20). This replication factor has been tested by x-ray scattering experiments by *Tidswell et al.* [365] for very thin cyclohexane films on rough silicon substrates (see Sects. 3.2.2 and 6.2.2 and Figs. 3.20 and 6.11). However, it was not possible to separate unambiguously the conformal part from the intrinsic part of the roughness.

A way to achieve this separation is by the use of periodic substrates, i.e. surface gratings [211, 212, 339, 371, 372, 373]. In the following the case of PS films (M_{W} = 1000 k) on top of surface gratings with spacings $\Lambda \approx$ 1 µm and heights $h \sim$ 150 Å is considered (see Fig. 6.18). Surface gratings may be regarded as a special kind of "roughness" with only a few enhanced Fourier components in the wavenumber spectrum. Hence, a systematic study of the aforementioned problem and a quantitative comparison with Eq. (6.25) become possible because the influence of the *microscopic statistical* part of the roughness[22] is now totally separated from the *periodic conformal* part.

If the situation of Fig. 6.18 is considered, then the continuous variable $q_\parallel$ has to be replaced by discrete values $q_m = m\, 2\pi/\Lambda$ (m integer) and the Fourier transforms in Eq. (5.50) have to be replaced by the respective Fourier coefficients, and thus

$$\tilde{z}_{\mathrm{l}}(q_m) = \chi(q_m)\, \tilde{z}_{\mathrm{s}}(q_m) = \frac{a^2}{a^2 + q_m^2\, d^4}\, \tilde{z}_{\mathrm{s}}(q_m)\ , \tag{6.26}$$

[22] The gratings themselves possess a microscopic roughness, and capillary wave fluctuations may be present on the surfaces of the liquid or polymer films.

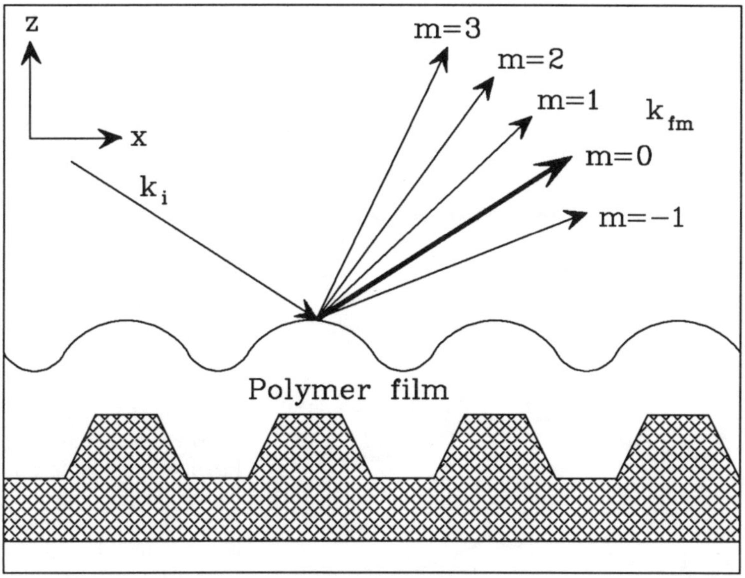

Fig. 6.18. Schematic drawing of a laterally structured surface with a polymer film on top. Owing to the periodicity, an incoming x-ray beam with wavevector $\boldsymbol{k}_\mathrm{i}$ is reflected ($m = 0$) and scattered into non specular directions ($m \neq 0$)

where $\tilde{z}_\mathrm{l}(q_m)$ and $\tilde{z}_\mathrm{s}(q_m)$ denote the mth Fourier coefficient of the liquid and the solid surface, respectively.

X-ray scattering data can be explained according to the theory outlined in Sect. 6.1. Here the extension to the particular case of periodic interfaces is briefly given in the kinematical approximation. A system of N layers on top of a surface described by $z_{\mathrm{s}_1}(x)$ with periodicity Λ (Si surface grating) is considered[23]. In practice, two overlayers, a native oxide layer $z_{\mathrm{s}_2}(x)$ and the polymer film $z_{\mathrm{l}_3}(x)$, have to be assumed[24].

The locations $z_k(\boldsymbol{r}_\parallel)$ of the interfaces are described by $z_k(\boldsymbol{r}_\parallel) = z_k(x) + \delta z_k(x) + d_k$, where $z_k(x)$ is assumed to be the periodic part of the kth interface $z_k(x) = z_k(x + \Lambda)$, $\delta z_k(x)$ is a random part (microscopic roughness) with vanishing mean value $\langle \delta z_k(x) \rangle_x = 0$, and d_k denotes the baseline of $z_k(x)$. With this notation the scattered x-ray intensity is

$$I(q_x, q_z) \sim \frac{\mathcal{G}}{q_z^4} \sum_m \sum_{j,k=1}^{N} \Delta\varrho_k \Delta\varrho_j \exp\left\{\mathrm{i}\, q_z(d_k - d_j)\right\} C^*_{k,m}(q_z)\, C_{j,m}(q_z)$$

$$\times \exp\left\{ - (\sigma_j^2 + \sigma_k^2) q_z^2/2 \right\} \left\{ 4\pi^2\, \delta(q_x - q_m) \right. \qquad (6.27)$$

[23] Only a one-dimensional lateral description is given.

[24] The index "s" means "solid" and "l" means liquid; the numbers denote the interface indices used in the equations.

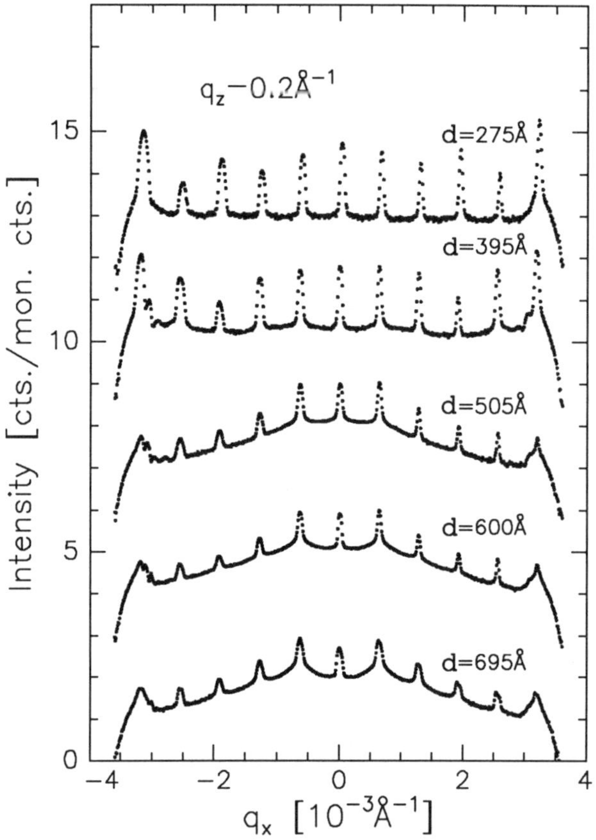

Fig. 6.19. Transverse scans of polystyrene layers on silicon gratings at $q_z = 0.2\,\text{Å}^{-1}$. The PS film thicknesses are $d = 275$–695 Å, the grating height is $h = 130$ Å, and the lateral period $\Lambda = 1\,\mu$m. The satellites next to the specularly reflected beam at $q_x = 0$ are located at $q_x = m\,2\pi/\Lambda$, with m being an integer. The separation between the conformal (narrow peaks) and the intrinsic nonconformal (scattering around the peaks) intensity is obvious. For large thicknesses the intrinsic part exhibits a power-law behavior, indicating long-range correlations typical for soft-matter surfaces

$$+\int\Big(\exp\{q_z^2 C_{jk}(X)\} - 1\Big)\exp\Big\{-\mathrm{i}\,(q_x - q_m)X\Big\}\,\mathrm{d}X\Big\}\,.$$

The asterisk denotes a complex conjugate quantity and the electron density differences of the layers are given by $\Delta\varrho_k = \varrho_{k+1} - \varrho_k$. As defined before in Chap. 5, $C_{jk}(X)$ is the height–height correlation function $C_{jk}(X) = \langle\delta z_j(x)\,\delta z_k(x+X)\rangle_x$ of the microscopic roughnesses between interfaces j and k, and $\sigma_k^2 = \langle[\delta z_k(x)]^2\rangle_x$ is the rms roughness of interface k. The quantities $C_{k,m}(q_z)$ are *not* the Fourier coefficients of the periodic functions $z_k(x)$ but those of the function $\exp\{-\mathrm{i}\,q_z z_k(x)\}$ (for a detailed discussion see Refs. [134, 163, 333, 367, 368, 370]).

The delta function $\delta(q_x - q_m)$ in Eq. (6.27) leads to resolution-limited diffraction orders at the positions $q_x = q_m = m\,2\pi/\Lambda$ in reciprocal space which are caused by the lateral periodicity of the system (see Fig. 6.18). The diffuse scattering from the random fluctuations $\delta z_k(x)$ is described by the integral in Eq. (6.27) according to the theory outlined in Sects. 6.1.1 and

6.1.2. Hence each diffraction order is riding on its "own" diffuse scattering. This can be seen in Fig. 6.19, where transverse scans for PS overlayers on a silicon surface grating are displayed. For the thicker films the power-law tails of the diffraction orders are the signature of the intrinsic roughness part $\delta z(\boldsymbol{r}_{\|})$ due to long-range correlations (see Sect. 6.1.3). The q_z dependence of the intensity of the diffraction orders contains the information about the periodic part of the interfaces[25].

Symmetric trapezoidally shaped surface gratings prepared by plasma-etching methods were used as substrates [154, 156, 371]. For these gratings a calculation of the coefficients $\mathcal{C}_{s_1,m}(q_z)$ can easily be done[26]. The substrates are covered with a perfectly conformal native oxide layer [$\mathcal{C}_{s_2,m}(q_z) = \mathcal{C}_{s_1,m}(q_z)$, $d_2 \approx 10\,\text{Å}$]. The baseline of the PS film is given by d_3, and σ_3 denotes the rms roughness of the PS/air interface.

The PS films were first spun onto a glass substrate. Estimates of their thicknesses were obtained by ellipsometry. Afterwards they were floated on a water surface. Finally, they were put on the surface of the gratings and then annealed for 2 h at 185°C in a vacuum oven. Before and after the x-ray experiments the surfaces of the deposited PS films were checked with a Digital Instruments Nanoscope III AFM.

The AFM measurements reveal that the contours of the adsorbed PS films $z_{l_3}(x)$ are almost sinusoidal. Therefore a Fourier expansion

$$z_{l_3}(x) = \sum_{n>0} \tilde{z}_{l_3}(q_n)\cos(q_n x) \;, \tag{6.28}$$

with only one dominant Fourier component, i.e. $|\tilde{z}_{l_3}(q_n)| \ll |\tilde{z}_{l_3}(q_1)|$ for $n \geq 2$, is straightforward and leads to the expression

$$\mathcal{C}_{l_3,m}(q_z) = \mathrm{i}^{-m}\mathrm{J}_m[q_z\tilde{z}_{l_3}(q_1)] - \frac{1}{2}q_z \sum_{n>1} \mathrm{i}^{n-m+1}\tilde{z}_{l_3}(q_n) \tag{6.29}$$

$$\Big\{\mathrm{J}_{m-n}[q_z\tilde{z}_{l_3}(q_1)] + (-1)^n \mathrm{J}_{m+n}[q_z\tilde{z}_{l_3}(q_1)]\Big\} \;,$$

where $\mathrm{J}_m(x)$ are the Bessel functions of integer order, for the PS overlayer.

In Fig. 6.20, scans along q_z for fixed $q_x = q_m$ with $m = 0$–4 are depicted. The measurements (open circles) and fits (solid lines) using Eq. (6.27) are shown ($d_3 = 275\,\text{Å}$). Three Fourier coefficients $\tilde{z}_{l_3}(q_1)$, $\tilde{z}_{l_3}(q_3)$, and $\tilde{z}_{l_3}(q_5)$ were refined[27]. It turns out that the fits are very sensitive to the coefficients $\tilde{z}_{l_3}(q_1)$ and $\tilde{z}_{l_3}(q_3)$, and less sensitive to $\tilde{z}_{l_3}(q_5)$. This means that a determination of $\chi(q_1)$ and $\chi(q_3)$ is possible.

Figure 6.21 shows the replication factor as obtained for various film thicknesses (symbols) [212, 371, 372]. The solid and dashed lines are fits of the

[25] This does not mean that the rms roughnesses σ_k can be neglected (see exponential in Eq. 6.27).

[26] An analytic expression is given in Refs. [367, 368].

[27] Since the widths of the bars and grooves are equal, only odd Fourier coefficients occur.

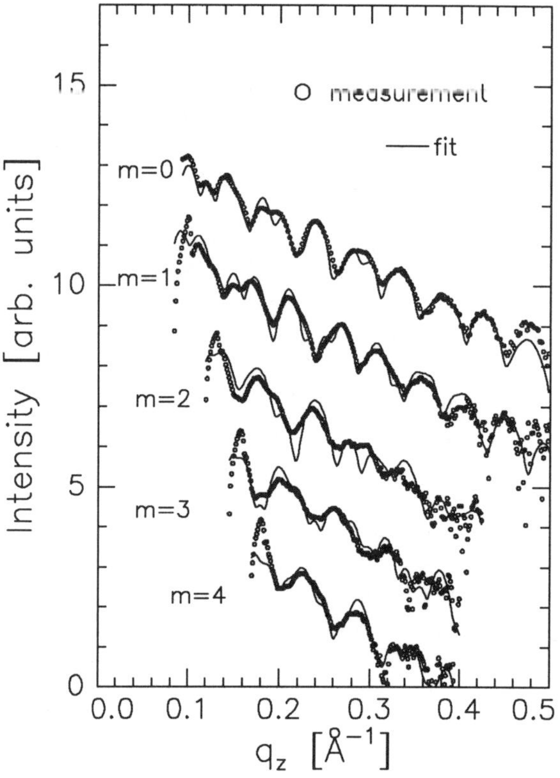

Fig. 6.20. X-ray-scattering data (λ = 1.13 Å, X10B beamline NSLS Brookhaven) from a d = 275 Å thick PS overlayer on top of a silicon surface grating with lateral period Λ = 1 μm. The specular reflectivity corresponds to the curve for $m = 0$. The other four curves are scans along q_z for fixed $q_x = m\,2\pi/\Lambda$. *Lines*: fits, from which three Fourier coefficients of the PS surface were obtained. The curves are shifted on the intensity scale for clarity; the diffuse scattering stemming from the random fluctuations $\delta z_k(x)$ was determined by q_z scans at the positions $q_x = m\,2\pi/\Lambda + \pi/2$ and subtracted afterwards [372]

data using Eq. (6.25) with $q_\parallel = q_1$ and $q_\parallel = q_3$, respectively. The inset shows the calculated structure of the system. The damping of the roughness as a function of the film thickness can be seen. The fit yields a value of $a = 75$ Å, whereas a value of only $a \approx$ 5–10 Å is typical for the system Si/SiO$_2$/PS [387, 413]. The difference is rather large, which leads one to suspect that either the principal assumption of pure van der Waals interactions between the PS film and the substrate may not be valid, or that the linear-response theory may break down for this particular system.

For polymer films in a glassy state the discrepancy between the measured value $a = 75$ Å and the theoretically predicted $a \approx$ 5–10 Å can be explained with a linear-response theory but using modified interactions. The excess free energy per unit area of a thin polymer film of thickness d, with n chains/unit area, $R_0 = \sqrt{6}\,R_\mathrm{G}$, and constant volume is [76, 413] (see also Eq. 5.18 in Sect. 5.3.3)

$$\Delta F = \gamma + \gamma_\mathrm{sf} - \gamma_\mathrm{sv} + \frac{A_\mathrm{eff}}{12\pi d^2} + \frac{\pi^2}{6} k_\mathrm{B} T n \left(\frac{R_0^2}{d^2} - 1 \right) . \tag{6.30}$$

Ordering the terms of Eq. (6.30) according to powers of d leads to

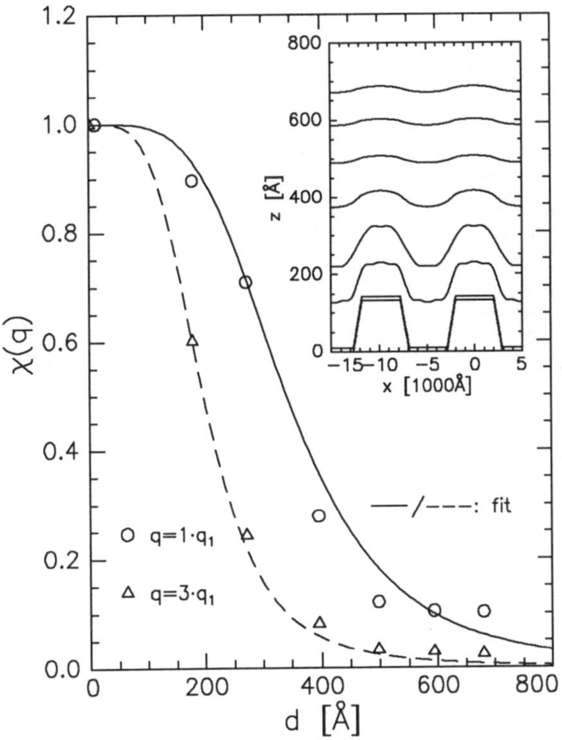

Fig. 6.21. Replication-factor $\chi = \tilde{z}_l/\tilde{z}_s$ obtained from the fits of the x-ray scattering data (see e.g. Fig. 6.20) for PS overlayers on a silicon grating with lateral period $\Lambda = 1\,\mu m$. *Symbols*: data. *Solid* and *dashed lines*: fits according to the model of *Andelmann et al.* [14] (see Eq. 6.26) for two wavenumbers $q_1 = 2\pi/\Lambda$ and $q_3 = 6\pi/\Lambda$. The inset shows the structure of the system in real space, directly visualizing how the substrate roughness is transferred through the polymer overlayers [372]

$$\Delta F = \hat{\gamma} + \gamma_{sf} - \gamma_{sv} + \frac{\hat{A}_{eff}}{12\pi d^2}\,, \tag{6.31}$$

where the definitions $\hat{\gamma} = \gamma - \pi^2 k_B T n/6$ and $\hat{A}_{eff} = A_{eff} + 2\pi^3 k_B T n R_0^2$ have been introduced. This equation is similar to the free energy of a liquid with surface tension $\hat{\gamma}$ and van der Waals-like interactions described by $\hat{A}_{eff}$. Therefore, in the theory of *Joanny, Andelmann & Robbins* [14, 298], γ and A_{eff} may be simply replaced by $\hat{\gamma}$ and $\hat{A}_{eff}$, and, hence, in Eq. (6.25) the length a may be replaced by $\hat{a} = \{\hat{A}_{eff}/(2\pi\hat{\gamma})\}^{1/2} \approx 100\,Å$. In the present case[28] this is close to the observed value of 75 Å.

Hence, this study has shown that polymers behave like liquids on rough surfaces but one has to take account of the fact that they consist of long molecules by means of the last term in Eq. (6.30), which is an entropic contribution to the free energy.

Another class of soft-matter films where the replication factor is theoretically known and has been tested experimentally is that of smectic ordered diblock copolymer films. Experimentally the properties of copolymers on flat

[28] $R_0 = 6.7\,Å\,\sqrt{N}$, where N is the number of monomers, $n = d/(163\,N)\,Å^{-3}$, $T \sim 300\,K$, and $d \sim 300\,Å$.

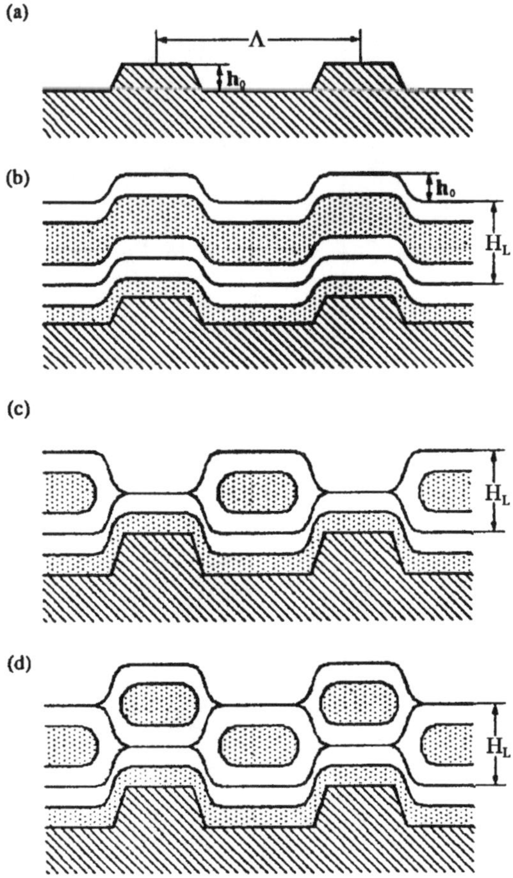

Fig. 6.22. Schematic drawing of PS–P2VP diblock copolymer configurations on patterned substrates. **(a)** Bare grating with period Λ and height ("roughness") h_0. **(b)** Conformal arrangement of a diblock copolymer lamella of height H_L on a grating with $h_0 < h_c$, where h_c is the "critical height". **(c)** Anticonformal configuration of a copolymer film on a grating with $h_0 > h_c$. **(d)** Arrangement of multiple layers in the anticonformal case. Note that the uppermost two layers are anticonformal to each other (see text). The shaded regions correspond to the P2VP component. A thin conformal P2VP layer next to the substrate is always present [211]

surfaces have been extensively studied (see e.g. Ref. [13]). It is well known that A–B symmetric diblock copolymers can be induced to form lamellar layers ordered parallel to the substrate surface. Islands or holes can be formed if the film thickness is not an exact multiple of the period, which is defined by the lamellar height H_L.

In 1992 *Turner & Joanny* [381] studied theoretically the ordering of symmetric diblock copolymers on a sinusoidal surface. Their premise was that the amplitude h_0 of the sinusoidal surface is small compared to the lamellar height. This study yielded a very surprising result: If the lateral period of the surface Λ is large compared to H_L, then the film surface contour is in phase with that of the substrate ("conformal arrangement", see Fig. 6.22b). On the other hand, if $\Lambda \ll H_L$, then the diblock copolymer surface is completely out of phase with respect to the underlying periodic substrate ("anticonformal arrangement", see Fig. 6.22c,d). This theoretical prediction is difficult to prove since most techniques only probe the topmost surface contour of a sample. Thus, the structures in Figs. 6.22b,c would appear as identical.

In 1997 *Li et al.* [211] showed with a detailed AFM study that anticonformal arrangements may indeed occur on periodic surfaces. However, a different situation from that investigated by *Turner & Joanny* [381] was studied. *Li et al.* [211] proved, both experimentally and theoretically, that a conformal/anticonformal transition happens as a function of the grating height h_0. If h_0 is smaller than a "critical height" $h_c \approx H_L/3$ then the structures are conformal; otherwise, i.e. for $h_0 > h_c$, the diblock copolymer films are anticonformal.

The ultimate answer to the question of whether an overlayer is conformal or anticonformal may be obtained by x-ray scattering investigations. With Eq. (6.27), the scattered intensity $I_m(q_z)$ for fixed $q_x = m\,2\pi/\Lambda$ can be easily obtained as a function of q_z. Assuming only one polymer overlayer of thickness d with exactly the same contour as the substrate grating, i.e. $\mathcal{C}_{m,1}(q_z) = \mathcal{C}_{m,2}(q_z) = \mathcal{C}_m(q_z)$ (see Eq. 6.27), and neglecting roughness leads to

$$I_m(q_z) \sim \frac{|\mathcal{C}_m(q_z)|^2}{q_z^4}\left\{(\Delta\varrho_1 + \Delta\varrho_2)^2 - 4\Delta\varrho_1\Delta\varrho_2 \sin^2(q_z d/2)\right\} \qquad (6.32)$$

in the conformal case, where $\Delta\varrho_1$ and $\Delta\varrho_2$ denote the density contrasts between the grating material and the polymer and at the polymer/air interface. The anticonformal arrangement, i.e. $\mathcal{C}_{m,1}(q_z) = -\mathcal{C}_{m,2}(q_z)$, yields

$$I_m(q_z) = \frac{|\mathcal{C}_m(q_z)|^2}{q_z^4}\left\{(\Delta\varrho_1 + \Delta\varrho_2)^2 - 4\Delta\varrho_1\Delta\varrho_2 \begin{matrix}\sin^2\\ \cos^2\end{matrix}\Big(q_z d/2\Big)\right\}, \qquad (6.33)$$

where the sine has to be taken when the integers m are even, including zero, and the cosine when m is odd.

Hence, the two arrangements of diblock copolymer films on surface gratings can be distinguished by measuring the intensities of the odd diffraction orders along q_z lines with constant $q_x = m\,2\pi/\Lambda$ in reciprocal space. If the thickness oscillations are in phase with those of the even orders then the structure is conformal (see Eq. 6.32); otherwise it is anticonformal (see Eq. 6.33). Figure 6.23 shows data[29] (symbols) and calculations according to Eqs. (6.32) and (6.33). In Fig. 6.23a the phase difference can be seen. Since some parts of the sample were covered with islands of the next layer, the phase shift is not $\pi/2$ but nevertheless is clearly visible[30]. Thus, the anticonformal arrangement has been proven by x-ray scattering, too.

It turns out that there is almost no roughness attenuation in the anticonformal case [211]. Hence the replication factor

$$\chi(q_x) \approx \begin{cases} 1 - H_L/h_0 & \text{for } n \text{ odd}, \\ 1 & \text{for } n \text{ even}, \end{cases} \qquad (6.34)$$

[29] Measured at HASYLAB (Hamburg) beamline E2 with $\lambda = 1.00$ Å.

[30] Expressions which are more complicated than those given by Eqs. (6.32) and (6.33) can be derived for this situation. It turns out that essentially the pure $\cos^2(q_z d/2)$ term has to be replaced by $\sin^2(q_z d/2 \pm \varphi)$ with $\varphi < \pi/2$.

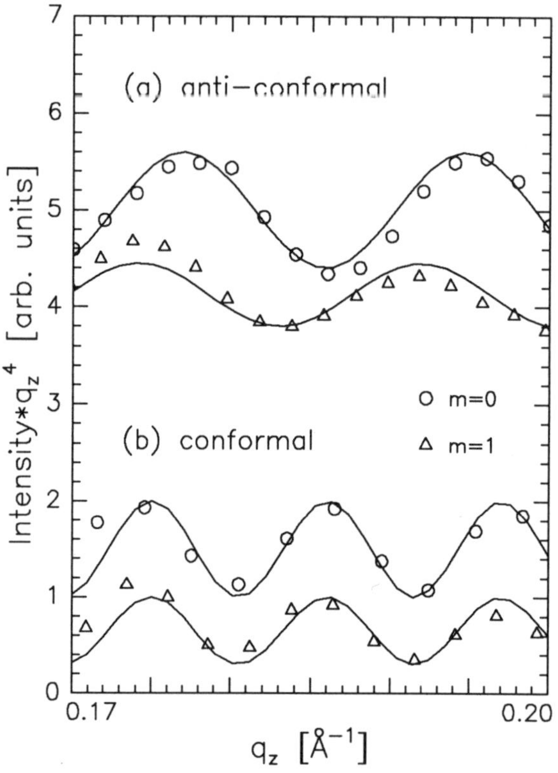

Fig. 6.23. Reflectivity normalized by the q_z^{-4} decrease of the Fresnel reflectivity (*open circles*, $m = 0$) and intensities of the first diffraction orders (*open triangles*, $m = 1$) of PS–P2VP block copolymer films on silicon surface gratings with period $\Lambda = 1\,\mu$m. The *lines* are fits. **(a)** Data for the anticonformal arrangement for an $H_L \approx d = 350\,$Å thick film on a grating with height $h_0 = 160\,$Å ($h_0 > H_L/3$). **(b)** Data for the conformal arrangement, $H_L \approx d = 580\,$Å, $h = 80\,$Å ($h_0 < H_L/3$). The anticonformality is proven by the different phases of the thickness oscillations

may be empirically deduced, where n denotes the number of overlayers of thickness $d \approx H_L$. Equation (6.34) takes into account the fact that alternate anticonformal/conformal configurations with respect to the grating are present (see Fig. 6.22d).

Finally, the replication factor for the conformal case of smectic ordered diblock copolymer films will be discussed more quantitatively. Minimization of the Landau–de Gennes free energy [74, 77] for a smectic film on a periodic substrate yields, in mean-field approximation (for details see *Li et al.* [211]),

$$\chi(q_x) = \frac{1}{\cosh(\xi\, q_x^2\, d) + \Gamma \sinh(\xi\, q_x^2\, d)} . \tag{6.35}$$

Here ξ is a length characteristic of the smectic film defined by $\xi = \sqrt{K/B}$, where B and K are the compressional and bending moduli. These moduli may be estimated from appropriate microscopic theories [382]. The length ξ is predicted to be on the order of the lamellar period H_L [74, 382]. In addition, $\Gamma = \gamma/(K\,B)^{1/2}$ has been introduced, with γ being the surface tension[31].

[31] For the system PS–P2VP a value of $\Gamma \approx 30$ is typical.

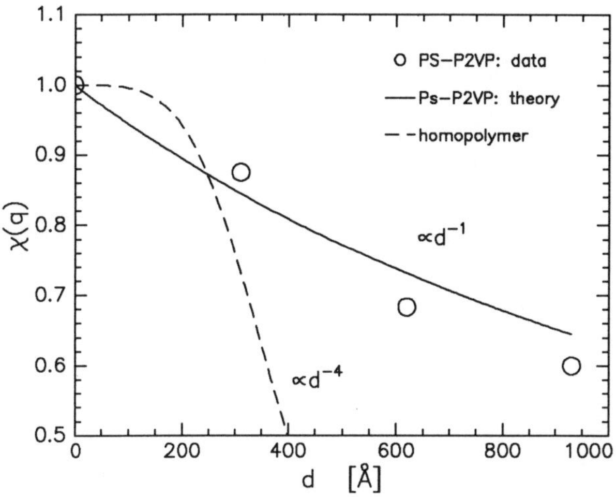

Fig. 6.24. Measured roughness attenuation $\chi(q_x)$ (*open circles*) and calculation (*solid line*) using a replication factor based on minimization of the free energy for a smectic diblock copolymer film. The *broken line* shows the rapidly decreasing $\chi(q_x)$ for a homopolymer [211]

Since the approximation $\xi\, q_x^2\, d \ll 1$ is valid for most systems, the denominator of Eq. (6.35) may be expanded. Thus

$$\chi(q_x) = \frac{1}{1 + q_x^2\, b\, d} \sim d^{-1}\,, \tag{6.36}$$

with $b = \Gamma\xi$. For small-molecule liquids and homopolymers (see Sect. 5.4) the replication factor is given by [14, 212, 365, 372]

$$\chi(q_x) = \frac{1}{1 + q_x^2 a^{-2}\, d^4} \sim d^{-4}\,, \tag{6.37}$$

where a is the "healing length", which is on the order of 10 Å for simple liquids and on the order of 100 Å for long-chain homopolymers (see the example above). Figure 6.24 shows measurements of $\chi(q_x)$ (open circles) as a function of the PS–P2VP overlayer thickness d together with a calculation according to Eq. (6.36). The broken line, valid for a homopolymer (Eq. 6.37), cannot be adjusted to the data.

Thus, Eqs. (6.36) and (6.37) indicate that vertical roughness propagation in thin diblock copolymer films is much more efficient ($\chi \propto d^{-1}$) than in homopolymer films ($\chi \propto d^{-4}$). This finding may be useful in thin-film technology, where perfectly conformal insulating overlayers are required in many practical applications.

Here the discussion of roughness propagation and that of the "usual" scattering experiments stops. The next chapter deals with a relatively young field: coherent x-ray scattering. This is a technique where the x-ray beam is sensitve to the actual contour function $z(\boldsymbol{r}_\parallel)$ of an interface rather than yielding information about statistical averages.

7. X-Ray Scattering with Coherent Radiation

In the previous chapters x-ray scattering from soft-matter surfaces was discussed in terms of a theory based on a statistical description of interfaces. This is justified since in a "normal" experiment the coherence volume of the impinging radiation can be assumed to be much smaller than the illuminated sample area and aperture dimensions. Hence, the theory outlined in Chaps. 2 and 6 may be more precisely termed the "incoherent x-ray scattering theory".

With the advent of high-brilliance third-generation synchrotron x-ray sources it is now possible to obtain intense x-ray beams possessing a high degree of coherence[1]. Several experiments have been carried out recently that demonstrate the coherence properties of such beams, such as Fraunhofer diffraction patterns [148, 214, 217, 388] (see Fig. 7.1), the observation of speckle patterns [55, 214, 299, 300, 361] and their fluctuations in time, and so-called x-ray photon correlation spectroscopy (XPCS) [87, 245, 364].

However, most experiments are in practice carried out with radiation that is only partially coherent, and a quantitative understanding of the observed diffraction or speckle patterns depends on a proper theory for incorporating the effects of partial coherence on the scattering. This chapter will provide insight into the results of such a theory beyond the usual relatively simplistic considerations based on the discussion of the quantities

$$\xi_{\mathrm{t}} = \frac{\lambda L}{s} \quad \text{and} \quad \xi_{\mathrm{l}} = \frac{\lambda^2}{\Delta\lambda} \, . \tag{7.1}$$

These quantities are the transverse and longitudinal coherence lengths ξ_{t} and ξ_{l}, respectively, where L denotes the distance between an (incoherent) source of spatial extent s and $\Delta\lambda/\lambda$ is the monochromaticity of the radiation [44, 103, 234, 369].

A comparison between incoherent scattering and scattering with coherent radiation is given in the next section. Afterwards, the general formalism is briefly discussed with emphasis on the particular case of surfaces, and finally a short introduction to XPCS is given. Details of all calculations can be found in Refs. [340, 375, 377] and Appendices A.3 and A.4.

[1] The coherent photon flux F_{c} of a source is given by $F_{\mathrm{c}} = (\lambda/2)^2(\Delta\lambda/\lambda)\mathcal{B}$, where $\mathcal{B}$ is the brilliance and $\Delta\lambda/\lambda$ is the monochromator bandpass [149, 300].

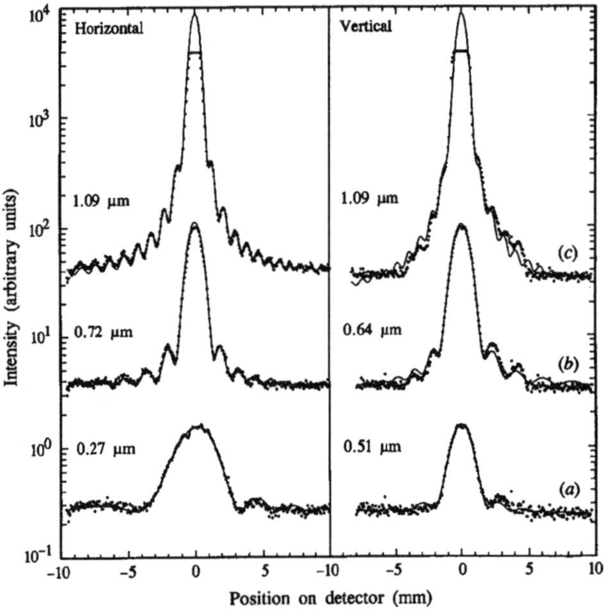

Fig. 7.1. Horizontal and vertical traces (*closed circles*: data; *lines*: fits) through Fraunhofer patterns from a square slit with horizontal and vertical dimensions **(a)** 0.27 μm × 0.51 μm, **(b)** 0.72 μm × 0.64 μm, and **(c)** 1.09 μm × 1.09 μm. The flat tops of the curves (c) obtained with the largest slit are due to saturation of the detector. The data were taken at the ESRF beamline ID3 (figure from *Vlieg et al.* [388])

7.1 Coherent versus Incoherent Scattering

Most treatments of the effects of finite beam divergence, energy spread, etc. use the so-called "resolution-function-folding procedure", i.e. the observed intensities are calculated in terms of a convolution of the actual scattering function $S(\boldsymbol{q})$ with an instrumental resolution function $\tilde{R}(\boldsymbol{q})$ as discussed in Sect. 3.1.1. This treatment is valid only in the limit when (i) the sample sees a completely incoherent source[2], and (ii) in the far-field (Fraunhofer) diffraction regime. It is quite simple to prove that the conditions for Fraunhofer diffraction are much more stringent for x-rays than for light, so that unless the aperture distances from the sample are very large, one must use Fresnel rather than Fraunhofer diffraction theory [44].

To show the principal difference between coherent and incoherent scattering a system described in terms of an electron density function $\varrho(\boldsymbol{r})$ placed

[2] Note that even radiation from such a source may develop a finite degree of coherence at the sample position, if the latter is sufficiently far away from the source!

in a perfectly monochromatic and collimated x-ray beam (single incident wavevector $\boldsymbol{k}_{\mathrm{i}}$) is considered. The cross section for scattering is given by

$$\frac{\mathrm{d}\sigma}{\mathrm{d}\Omega} = P_1 r_{\mathrm{e}}^2\, S_T(\boldsymbol{q}), \tag{7.2}$$

where r_{e} is the Thomson scattering length of the electron (see Sect. 2.1), P_1 is the Lorentz polarization factor, and

$$S_T(\boldsymbol{q}) = \iint \langle \varrho(\boldsymbol{r})\varrho(\boldsymbol{r}')\rangle_T \exp\left\{\mathrm{i}\,\boldsymbol{q}\cdot(\boldsymbol{r}'-\boldsymbol{r})\right\} \mathrm{d}\boldsymbol{r}\mathrm{d}\boldsymbol{r}' \tag{7.3}$$

is the kinematical scattering function as defined by Eq. (6.2) in Sect. 6.1.1. Here it is also assumed that the measurement is made over a long enough period of time T that a time average of the correlation function inside the integral can be performed.

It should be noted that no statistical ensemble averages in the above equations have been taken, since a particular realization of a sample in a perfectly coherent beam is considered. If the sample is non-ergodic, i.e. has built-in static randomness or disorder, $S_T(\boldsymbol{q})$ will possess sharp and random fluctuations about some particular average function, giving rise to the phenomenon of "speckle". If, on the other hand, the system is ergodic, with fluctuation timescales very short compared to the total integrated counting time T (e.g. a normal liquid), the time average is equivalent to an ensemble average and $S_T(\boldsymbol{q})$ may be replaced by the usual ensemble average. In this case, even for completely coherent radiation, there is no speckle. Instead, what is observed is a smooth function $S(\boldsymbol{q})$ that can be calculated in the usual way by statistical mechanical methods.

For conventional scattering experiments, the diffuse scattering from disordered solids, rough surfaces, etc. does not usually exhibit speckle but instead gives a smooth $S(\boldsymbol{q})$, which is in accordance with an ensemble average, notwithstanding the fact that the disorder in such systems is non-ergodic. There are two ways to understand this. The first is a "resolution function" smearing of the speckle pattern in reciprocal space as described in Sect. 3.1.1. In this picture, Eq. (7.2) is folded with an instrumental resolution function as if a whole series of beams with a distribution of energies and incident and scattered directions independently, i.e. incoherently, scatter from the sample. Mathematically, this procedure is described by Eqs. (3.8) and (3.9) in Sect. 3.1.1.

The second way to understand the smoothing is in terms of finite coherence volumes for the radiation in the sample. It is assumed that these are centered at positions $\boldsymbol{R}_l$ throughout the volume of the sample and are defined in terms of some (three-dimensional) real-space cutoff function $P(\boldsymbol{r})$. The intensity $I_l = |\dots|^2$ of the scattering from a single volume centered at

$\boldsymbol{R}_l$ must be added over all such volumes, since they scatter incoherently from each other, yielding the total intensity[3]

$$I(\boldsymbol{q}) = \sum_l \left| \int \varrho(\boldsymbol{r}) P(\boldsymbol{r} - \boldsymbol{R}_l) \exp(-\mathrm{i}\,\boldsymbol{q}\cdot\boldsymbol{r})\, \mathrm{d}\boldsymbol{r} \right|^2 . \tag{7.4}$$

With the Fourier transforms of $\varrho(\boldsymbol{r})$ and $P(\boldsymbol{r})$ and assuming a large enough set of $\boldsymbol{R}_l$, Eq. (7.4) is equivalent to

$$I(\boldsymbol{q}) = \int \left|\tilde{\varrho}(\boldsymbol{K})\right|^2 \left|\tilde{P}(\boldsymbol{q}-\boldsymbol{K})\right|^2 \mathrm{d}\boldsymbol{K} , \tag{7.5}$$

which is identical to the form of Eq. (3.8) if $|\tilde{P}(\boldsymbol{K})|^2$ is identified with the "resolution function" $\tilde{R}(\boldsymbol{K})$. Thus, in this incoherent limit[4] the conventional method of simply folding the true $S(\boldsymbol{q}) = |\tilde{\varrho}(\boldsymbol{K})|^2$ with a resolution function in reciprocal space becomes correct, and the resultant scattering provides a reasonable approximation to an ensemble average, even though the experiment has involved a single realization of the sample.

Equation (7.5) has traditionally been used to obtain statistically averaged $S(\boldsymbol{q})$ functions for the system studied by "unfolding" the resolution function or by fitting. However, most x-ray radiation, whether emitted from an incoherent source or not, possesses a finite degree of coherence by the time it is incident on the sample. Hence, it is worth re-analyzing the expression for the observed intensity along the lines that researchers in optics have followed. The calculations of the next section show (see also Ref. [340]) that Eq. (3.9) in Sect. 3.1.1 must be replaced by

$$I(\boldsymbol{q}) = \iint \varrho_{\mathrm{F}}(\boldsymbol{r})\varrho_{\mathrm{F}}^*(\boldsymbol{r}')\, R(\boldsymbol{r}',\boldsymbol{r}) \exp\left\{\mathrm{i}\,\boldsymbol{q}\cdot(\boldsymbol{r}'-\boldsymbol{r})\right\} \mathrm{d}\boldsymbol{r}\, \mathrm{d}\boldsymbol{r}' , \tag{7.6}$$

where $\varrho_{\mathrm{F}}(\boldsymbol{r})$ is *not* the physical electron density but a modified "Fresnel electron density" which is $\varrho(\boldsymbol{r})$ multiplied by a (complex) phase factor that depends on the directions of the incident and scattered beams, and the function $R(\boldsymbol{r}',\boldsymbol{r})$ does *not* depend only on the separation between $\boldsymbol{r}'$ and $\boldsymbol{r}$.

In general, this considerably complicates the interpretation of scattering data. For highly coherent beams, $R(\boldsymbol{r}',\boldsymbol{r})$ contains "slit diffraction effects" and $\varrho_{\mathrm{F}}(\boldsymbol{r})$ contains "finite-sample diffraction effects" that may yield highly oscillating forms for $I(\boldsymbol{q})$. A more detailed, rigorous calculation of the scattering is given in the next section.

7.2 General Formalism

In practice, beams of radiation falling on samples have encountered several optical elements on their way from the source, e.g. monochromators, mirrors,

[3] For clarity the Thomson scattering and the Lorentz factors are omitted in the following.

[4] The coherence volume is much smaller than the sample volume.

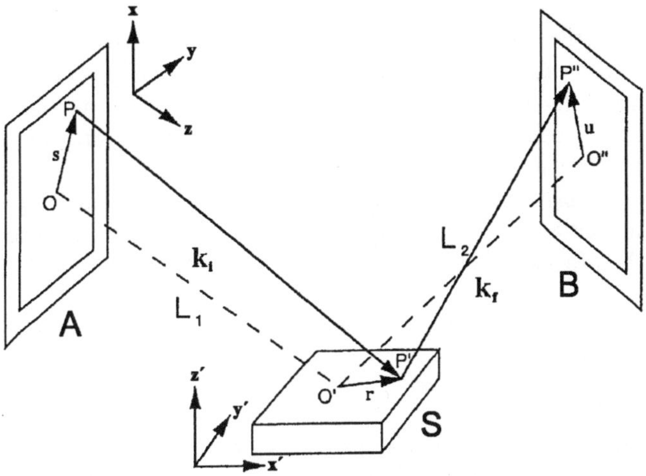

Fig. 7.2. Schematic view of the scattering geometry. The laboratory-fixed coordinate system is given by (x, y, z), and (x', y', z') is the more convenient sample-fixed system. The mean incident and outgoing beam directions are $\boldsymbol{k}_{\mathrm{i}}$ and $\boldsymbol{k}_{\mathrm{f}}$. Two traces of x-rays, $PP'P''$ and $OO'O''$, originating from aperture A are shown

slits, etc., and thus it is often more useful to work in terms of the statistical properties, i.e. the mutual coherence function (MCF) of the radiation[5] across the last aperture before the sample, and calculate how the MCF is propagated via the scattering across an "outgoing beam" aperture, and also the intensity in a detector behind this last aperture. This scenario is sketched in Fig. 7.2.

Two major approximations are made in the following: (i) The distances L_1, L_2 (see Fig. 7.2) from the sample to the incoming and outgoing apertures A and B, respectively, are sufficiently large compared to the sample and aperture dimensions. Hence only terms up to second order need to be considered in the ratios of the latter to the former. The calculations are valid in both the Fraunhofer and Fresnel regimes for diffraction[6]. The precise conditions for these regimes are discussed in detail in Refs. [340, 375, 377]. Here only the basic formulas which are needed for a data analysis will be given. (ii) The scattering is calculated in the kinematical approximation[7] (see Sects. 4.1 and 6.1.1).

In Fig. 7.2 the beam emerges from an aperture A (slit), the plane of which is normal to the mean direction $\boldsymbol{k}_{\mathrm{i}}$ of the beam and the line joining its center to the sample center a distance L_1 away; it is then transmitted through an aperture B oriented normal to the average direction $\boldsymbol{k}_{\mathrm{f}}$ of the scattered beam

[5] For details concerning the MCF see the textbooks of *Born & Wolf* [44] and *Mandel & Wolf* [234].

[6] The theory is also valid in the limit $L_1 \to 0$ with minor corrections [340].

[7] An extension to dynamical scattering is – in principle – possible. This has not been done yet.

at a distance L_2 from the center of the sample S and, finally, is counted in a detector D behind aperture B. According to the standard Huygens–Fresnel principle [44], the statistical properties of the electric field at the aperture B and the intensity in the detector D can be completely specified if one knows the spatial and temporal behavior of the electric field across aperture A and the electron density $\varrho(\boldsymbol{r})$ in the sample. For the moment time-dependent effects are neglected and $\varrho(\boldsymbol{r})$ is assumed to be a static function. The resultant scattering will be affected by both "coherence" and "resolution" effects, in the sense discussed in the previous section, but in general in a more complicated way.

The electric field $E(\boldsymbol{s},t)$ at position $\boldsymbol{s}$ in the aperture A measured relative to the center of A is given by[8]

$$E(\boldsymbol{s},t) = A(\boldsymbol{s},t)\exp(-\mathrm{i}\,\bar{\omega}t)\,, \tag{7.7}$$

where $\bar{\omega}$ is the average frequency of the radiation, and the time dependence of the amplitude $A(\boldsymbol{s},t)$ represents the relatively slow variation on timescales much larger than $1/\bar{\omega}$ due to the nonmonochromaticity of the beam. An MCF for the radiation is defined by

$$\Gamma(\boldsymbol{s},\boldsymbol{s}',\tau) = \langle A(\boldsymbol{s},t)A^*(\boldsymbol{s}',t+\tau)\rangle\;, \tag{7.8}$$

where the average represents a time average over t over many phase fluctuations of the radiation[9]. The following ansatz may be used as a rather general form for the MCF [44, 234, 265, 266]:

$$\Gamma(\boldsymbol{s},\boldsymbol{s}',\tau) = \Psi(\boldsymbol{s})\Psi^*(\boldsymbol{s}')\,g(\boldsymbol{s}-\boldsymbol{s}')\,F(\tau)\;I/A\,, \tag{7.9}$$

where I is the total beam intensity through the aperture area A. The form chosen for the MCF in Eq. (7.9) is not the most general possible form, but is commonly used and is known as the generalized Schell form [234, 322]. The function $\Psi(\boldsymbol{s})$ is called the amplitude factor, and $g(\boldsymbol{s}-\boldsymbol{s}')$ is called the coherence factor. The latter is defined to be unity when $\boldsymbol{s}=\boldsymbol{s}'$. The intensity of radiation at the position $\boldsymbol{s}$ in the slit is given by setting $\boldsymbol{s}=\boldsymbol{s}'$ and $\tau=0$ in Eq. (7.9), i.e. it is simply $|\Psi(\boldsymbol{s})|^2$. It is convenient to include the aperture cutoff function in the definition of $\Psi(\boldsymbol{s})$. The exact form of $\Psi(\boldsymbol{s})$ and $g(\boldsymbol{x})$ will depend on the nature and distance of the source of radiation from this aperture, and on the optical elements in the beam prior to the slit (mirrors, monochromators, etc.) and can be difficult to calculate. Thus the form of the MCF across the incident aperture is regarded as empirical, ultimately obtainable from experiment. However, it is quite useful to introduce here the concept of coherence lengths, which are implicitly contained in the MCF. If $g(\boldsymbol{x})$ is approximated by a Gaussian form, i.e.

$$g(\boldsymbol{x}) = \exp\Big\{-x^2/(2\,\xi_x^2)\Big\}\exp\Big\{-y^2/(2\,\xi_y^2)\Big\}\,, \tag{7.10}$$

[8] Polarization effects are neglected. They may easily be included.

[9] Typically, the phase fluctuations have a timescale of 10^{-15}–10^{-16} seconds for x-rays.

then ξ_x, ξ_y can be considered as the two transverse coherence lengths. The time autocorrelation function $F(\tau)$ decays with a characteristic time τ_l and a longitudinal coherence length may be defined by $\xi_l = c\,\tau_l$, c being the speed of light. It should be noted that since $\Gamma(\boldsymbol{s}, \boldsymbol{s}', \tau)$ has to obey Helmholtz equations in $\boldsymbol{s}$ and $\boldsymbol{s}'$, the ansatz given by Eq. (7.9) restricts the time autocorrelation function $F(\tau)$ to a pure exponential $F(\tau) = F_0 \exp(-\tau/\tau_l)$, where $F_0 = 1$ is set by definition [44, 234].

Using the Huygens–Fresnel principle [44] and assuming kinematical scattering (i.e. no multiple scattering from the sample) yields for the scattered intensity in a detector D placed behind aperture B [340]

$$I(\boldsymbol{q}) = \frac{r_e^2 \bar{\omega}}{2\pi\lambda^2} \frac{1}{L_1^2} \frac{1}{L_2^2} \frac{I}{A} S(\boldsymbol{q}) \,, \tag{7.11}$$

where λ is the mean wavelength of the radiation and the scattering function $S(\boldsymbol{q})$ is given by

$$S(\boldsymbol{q}) = \frac{q\,\xi_l}{k_{L_2}^4 c} \int \frac{1}{1 + (k\xi_l/q)^2\,(q-Q)^2} \left(\int_{\mathrm{B}} \hat{S}_Q(\boldsymbol{q} + k_{L_2}^2 \boldsymbol{u})\,\mathrm{d}\boldsymbol{u} \right) \frac{\mathrm{d}Q}{Q^2}, \tag{7.12}$$

where $k = 2\pi/\lambda$ is the mean k value of the impinging radiation, and $k_{L_1} = \sqrt{k/L_1}$ and $k_{L_2} = \sqrt{k/L_2}$ are introduced for notational convenience. The integral over $\mathrm{d}\boldsymbol{u}$ is taken over the outgoing aperture dimensions and the subscript Q means that $\hat{S}_Q(\boldsymbol{K})$ depends parametrically on Q. It should be emphasized that all components of the vectors are given in the laboratory-fixed coordinate system (x, y, z) as defined in Fig. 7.2.

Equation (7.12) has a simple interpretation. It is the function $\hat{S}_Q(\boldsymbol{K})$ folded with the resolution over the outgoing detector slits, and the resultant function which depends parametrically on Q folded with the Lorentzian resolution function for the longitudinal coherence centered at q. Note from Eq. (7.12) that the effective longitudinal coherence length for a particular experiment is ξ_l multiplied by the factor $k/q = 1/\sin(\Phi/2)$, thus magnifying this quantity considerably for experiments where q is small (e.g. surface scattering experiments) [103, 307, 369].

Equation (7.12) would be identical to the conventional formalism if $\hat{S}_Q(\boldsymbol{K})$ were the conventional function $S(\boldsymbol{K})$ folded with the angular resolution of the incident beam. However, $\hat{S}_Q(\boldsymbol{K})$ has a slightly more complicated form, namely[10]

$$\hat{S}(\boldsymbol{K}) = \iint \varrho_{\mathrm{F}}(\boldsymbol{r}) \varrho_{\mathrm{F}}^*(\boldsymbol{r}')\, R(\boldsymbol{r}', \boldsymbol{r}) \exp\left\{ \mathrm{i}\,(Q/q)\, \boldsymbol{K} \cdot (\boldsymbol{r}' - \boldsymbol{r}) \right\} \mathrm{d}\boldsymbol{r}\, \mathrm{d}\boldsymbol{r}' \,, \tag{7.13}$$

where the function $R(\boldsymbol{r}', \boldsymbol{r})$ is given by

[10] The subscript Q in $\hat{S}_Q(\boldsymbol{K})$ is omitted in the following.

$$R(\boldsymbol{r}',\boldsymbol{r}) = \iint \Psi(\boldsymbol{s})\Psi^*(\boldsymbol{s}')\, g(\boldsymbol{s}-\boldsymbol{s}') \exp\left\{\mathrm{i}\,(Q/q)\, k_{L_1}^2 (s^2 - s'^2)/2\right\}$$
$$\times \exp\left\{\mathrm{i}\,(Q/q)\, k_{L_1}^2 (\boldsymbol{s}'\cdot\boldsymbol{r}' - \boldsymbol{s}\cdot\boldsymbol{r})\right\} \mathrm{d}\boldsymbol{s}\,\mathrm{d}\boldsymbol{s}' . \quad (7.14)$$

Hence, the function $R(\boldsymbol{r}',\boldsymbol{r})$ is only a pure function of the distance $(\boldsymbol{r}'-\boldsymbol{r})$ if the coherence factor is $g(\boldsymbol{s}-\boldsymbol{s}') = \delta_{\boldsymbol{s},\boldsymbol{s}'}$, i.e. in the incoherent case $\xi_x, \xi_y \to 0$ (see Eq. 7.10). Then Eq. (7.13) is (almost) equivalent to the conventional resolution-folding procedure as described in Sect. 3.1.1. But even though $g(\boldsymbol{s}-\boldsymbol{s}') = \delta_{\boldsymbol{s},\boldsymbol{s}'}$, Eq. (7.13) is slightly different from the "normal" $S(\boldsymbol{q})$ given by Eq. (3.8) since the electron density is marked by a subscript F. This subscript indicates a modified electron density, referred to as the "Fresnel electron density" in the following, defined by[11]

$$\varrho_{\mathrm{F}}(\boldsymbol{r}) = \varrho(\boldsymbol{r}) \exp\left\{\mathrm{i}\,\frac{Q}{2q}\left(k_{L_1}^2 r_{\perp,1}^2 + k_{L_2}^2 r_{\perp,2}^2\right)\right\} , \quad (7.15)$$

where $\boldsymbol{r}_{\perp,1}$ and $\boldsymbol{r}_{\perp,2}$ are the components of $\boldsymbol{r}$ perpendicular to $\boldsymbol{k}_{\mathrm{i}}$ and $\boldsymbol{k}_{\mathrm{f}}$. Since k_{L_1} and k_{L_2} are very small for large L_1 and L_2, the Fresnel density $\varrho_{\mathrm{F}}(\boldsymbol{r})$ reduces to the true electron density $\varrho(\boldsymbol{r})$ only in this case. It is worth noting that the phase factor in Eq. (7.15), which originates from the inclusion of second-order terms in the calculations, considerably complicates the calculations of the scattering even in the kinematical limit (see Ref. [340]).

It is often preferable to evaluate the expressions for the scattered intensities in reciprocal rather than real space. By Fourier-transforming Eqs. (7.13) and (7.14) one may write $\hat{S}(\boldsymbol{K})$ in the reciprocal-space form

$$\hat{S}(\boldsymbol{K}) = \iint \Psi(\boldsymbol{s})\Psi^*(\boldsymbol{s}')\, g(\boldsymbol{s}-\boldsymbol{s}') \exp\left\{\mathrm{i}\,\frac{\Omega}{2} k_{L_1}^2 \left(s^2 - s'^2\right)\right\}$$
$$\times \tilde{\varrho}_{\mathrm{F}}\left(\boldsymbol{K} + \Omega k_{L_1}^2 \boldsymbol{s}\right) \tilde{\varrho}_{\mathrm{F}}^*\left(\boldsymbol{K} + \Omega k_{L_1}^2 \boldsymbol{s}'\right) \mathrm{d}\boldsymbol{s}\,\mathrm{d}\boldsymbol{s}' , \quad (7.16)$$

where the quantity $\Omega = Q/q = 1 + \Delta\lambda/\lambda$ has been introduced, with $\Delta\lambda/\lambda$ being the monochromaticity of the incoming radiation[12]. The arguments of the Fourier transform of the Fresnel density $\tilde{\varrho}_{\mathrm{F}}(\boldsymbol{K})$ represent the Cartesian components (K_x, K_y, K_z) relative to the (x,y,z) axes shown in Fig. 7.2, where the z axis is along the average incident beam direction. In scattering experiments sample-fixed coordinates are often more appropriate and a back transformation of the components has to be performed (see below).

Again, in the "incoherent limit" $(\xi_x, \xi_y \to 0)$, Eq. (7.16) reduces to

$$\hat{S}(\boldsymbol{K}) = \int \left|\Psi(\boldsymbol{s})\right|^2 \left|\tilde{\varrho}_{\mathrm{F}}\left(\boldsymbol{K} + \Omega k_{L_1}^2 \boldsymbol{s}\right)\right|^2 \mathrm{d}\boldsymbol{s} , \quad (7.17)$$

[11] Calculations for the scattering including second order terms, which also have led to a modified electron density quite similar to that presented here, have been carried out by *Durbin* [104] in connection with diffraction of a curved wavefront by a crystal.

[12] A perfectly monochromatic beam corresponds to $\Omega \equiv 1$.

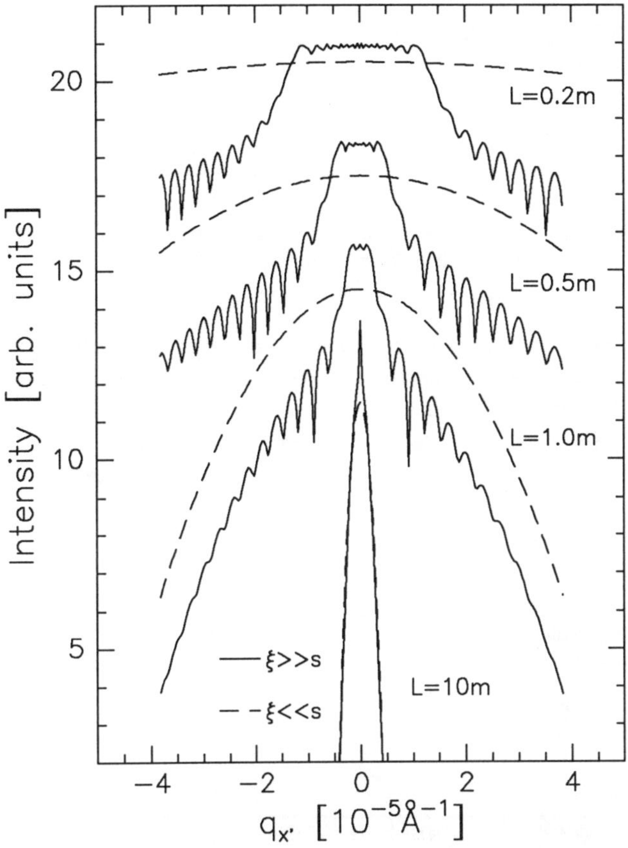

Fig. 7.3. Calculations of transverse $q_{x'}$ scans of the specularly reflected intensity from a single smooth surface in the coherent limit $\xi \gg s$ (*solid lines*) and the incoherent limit $\xi \ll s$ (*broken lines*) for various incident-slit/sample/exit-slit distances $L_1 = L_2 = L$ and the parameters $s = 50\,\mu$m (slit width), $\lambda = 1$ Å, $l = 1$ cm (sample size), and $q_{z'} = 0.44$ Å^{-1} ($\Phi = 2.0°$). Further explanations are given in the text [375]

i.e. $|\tilde{\varrho}_{\mathrm{F}}(\boldsymbol{K})|^2$ folded with the incident-aperture resolution function, as discussed above. In this case no incident-slit diffraction effects are expected, while in the "coherent limit" $g(\boldsymbol{s} - \boldsymbol{s}') \approx 1$ (i.e. ξ_x, ξ_y are much larger than the aperture dimensions) Eq. (7.16) may be written as

$$\hat{S}(\boldsymbol{K}) = \left| \int \Psi(\boldsymbol{s}) \exp(\mathrm{i}\,\Omega k_{L_1}^2 s^2/2)\, \tilde{\varrho}_{\mathrm{F}}\Big(\boldsymbol{K} + \Omega k_{L_1}^2 \boldsymbol{s}\Big)\, \mathrm{d}\boldsymbol{s} \right|^2 , \tag{7.18}$$

i.e. $\tilde{\varrho}_{\mathrm{F}}(\boldsymbol{K})$ first folded into the incident-aperture resolution including Fresnel diffraction effects, and then modulus squared. This can result in interference between different diffraction peaks which overlap in reciprocal space, resulting in complex diffraction patterns.

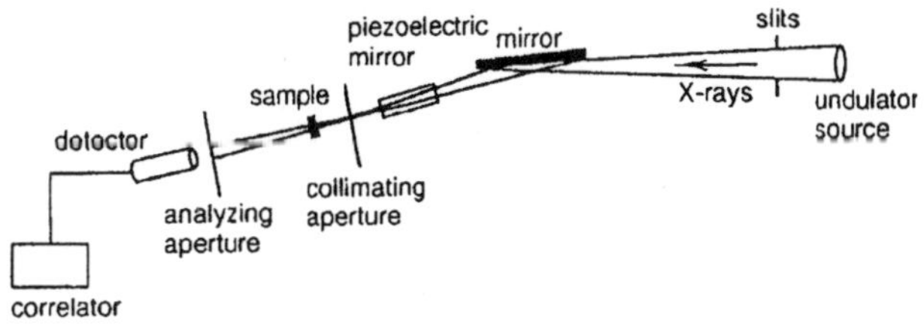

Fig. 7.4. Setup at the ESRF Troika beamline for measurements with coherent radiation. The correlator is needed for XPCS experiments [364]

As an example, the scattering from a single smooth surface is shown in Fig. 7.3. The calculations are outlined in the next section. Transverse $q_{x'}$ scans of the specularly reflected beam are depicted in the limits $\xi_x = \xi \gg s$ (solid lines) and $\xi \ll s$ (broken lines) for various incident-slit/sample/exit-slit distances $L_1 = L_2 = L$ (slit width $s = 50\,\mu\text{m}$). The quantity Ω was always set to unity and the detector-resolution folding in Eq. (7.12) was neglected. Figure 7.3 clearly shows the slit diffraction effects, and the reflected peak gets broader with decreasing distance L. The slit diffraction effects begin to disappear for larger distances and for $L = 10\,\text{m}$ the $\xi \gg s$ and $\xi \ll s$ cases are almost indistinguishable[13].

In the next section the coherent scattering from surfaces is considered in some more detail. Explicit expressions for $\tilde{\varrho}_{\text{F}}(\boldsymbol{K}')$ will be derived and compared qualitatively with experiments.

7.3 Coherent Scattering from Surfaces

In Chaps. 2 and 6 scattering from surfaces and interfaces was discussed. Implicitly, incoherent scattering was always assumed and statistical averages were performed on the way from Eq. (6.2) to Eq. (6.4), which finally led to a separation into a delta-like "specular" contribution and a diffuse component (see Eqs. 6.5–6.7). In the case of coherent scattering, the impinging radiation possesses at *all* points of the surface a fixed phase relationship. Thus, the radiation which is scattered from the sample contains information about the height contour $z(x, y)$ itself, rather than only statistical information via the correlation function $C(\boldsymbol{R}) = \langle z(\boldsymbol{r}_\parallel) z(\boldsymbol{r}_\parallel + \boldsymbol{R}) \rangle_{\boldsymbol{r}_\parallel}$ (see Chap. 5). Therefore, a separation into "specular" and "diffuse" scattering is not possible any more. Coherent scattering is both: It is somehow "specular" since it contains sharp, resolution-limited "speckle peaks" but may be considered as "diffuse" since it can also be found in directions other than the specular even for smooth surfaces (see Fig. 7.3).

[13] The coherence length at the sample is $\xi' = \lambda L/s \approx 20\mu\text{m} \sim s$ for the incoherent case, and hence the two cases are almost identical.

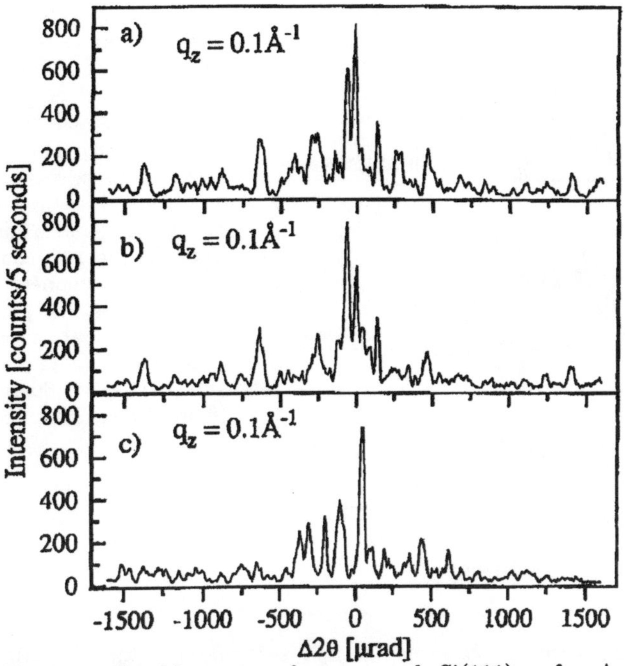

Fig. 7.5. Speckle pattern from a rough Si(111) wafer. A scan with an analyzer pinhole in the $\Delta 2\theta$ direction is shown ($\Delta 2\theta = \alpha_f - \alpha_i$ is the deviation from the specular condition in the scattering plane). The curves **(a)**, **(b)**, and **(c)** were recorded at different positions on the surface of the wafer (figure taken from *Libbert et al.* [215])

A typical experimental setup for coherent x-ray scattering requires pinholes of the size of microns, which may be used as collimating and analyzing apertures[14]. Figure 7.4 shows schematically such an experimental setup. Alternatively, a CCD camera may be used as an area detector instead of the analyzing pinhole. An extremely precise movement of these pinholes must be guaranteed in order to achieve the required wavevector transfers according to the equations given in Sect. 3.1.

Figures 7.5 and 7.6 show diffraction patterns which were recorded with a coherent x-ray beam. The data presented in Fig. 7.5 were obtained from a rough Si(111) wafer. Three speckle patterns were measured using 8 keV radiation from a synchrotron wiggler beamline (X25, NSLS Brookhaven). Different locations on the surface were illuminated, where the sample was only moved 20% along the beam direction between Figs. 7.5a and 7.5b. The pattern in fig. 7.5c was recorded at a different place. These scans demonstrate that the coherently scattered intensity is sensitive to the actual height function $z(x, y)$.

[14] Apertures A and B in Fig. 7.2.

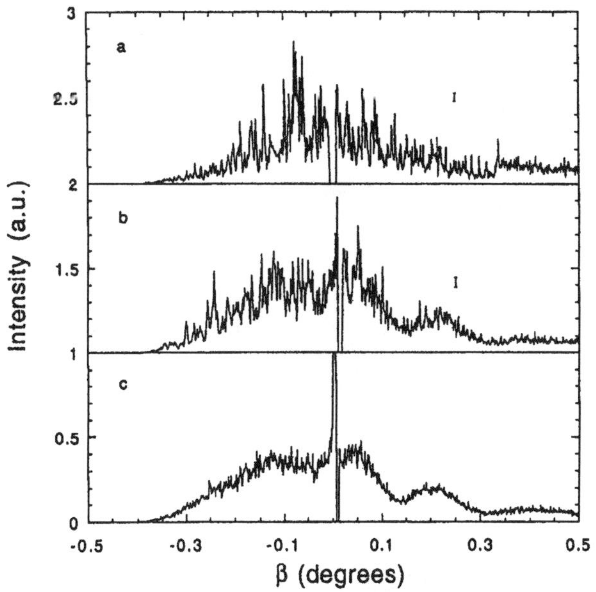

Fig. 7.6. Speckle patterns from an annealed thin film of diblock copolymers of PS and PMMA (coated with 200 Å of gold). The incident angle was $\alpha_i = 0.4°$ and $\beta = \alpha_f - \alpha_i$ is the deviation from the specular condition. The diameters of the collimating pinholes were **(a)** $s = 5\,\mu m$, **(b)** $s = 12\,\mu m$, and **(c)** $s = 60\,\mu m$ (figure taken from *Cai et al.* [55])

Figure 7.6 depicts speckle patterns from a soft-matter thin film observed with 6 keV radiation at a bending-magnet beamline (X6B, NSLS Brookhaven). An annealed diblock copolymer PS/PMMA film that consisted of micron-sized islands on a uniform surface was used as a sample. The incident angle was fixed at $\alpha_i = 0.4°$ during the scan[15] and the data were recorded with a CCD camera, from which line profiles are shown[16]. The angle $\beta = \alpha_f - \alpha_i$ is the deviation of the direction of the outgoing radiation from the specular condition. It can be seen that as the beam is made more coherent, by using the small $s = 5\,\mu m$ collimation pinhole, the speckle patterns show a higher visibility (see Fig. 7.6a) than those obtained with the larger pinholes (see Fig. 7.6b,c). For the $s = 60\,\mu m$ pinhole (Fig. 7.6c) the speckles almost disappear and only broad diffuse-scattering features can be seen. This example demonstrates the "smooth" transition from "fully coherent" to "incoherent" scattering, because Fig. 7.6b depicts an intermediate state: Speckle patterns as well as the diffuse, broad features are visible in the same scan.

However, as already mentioned at the beginning of Chap. 5, the seemingly nice fact that coherent scattering is sensitive to the ultimately obtainable information $z(x, y)$ is a disadvantage, too. The huge amount of information contained in $z(x, y)$ is often not of interest. In the case of coherent x-ray scattering this also means that it is quite difficult to compare the recorded

[15] This is well below the critical angle since the film was covered with 200 Å of gold.

[16] The intensity in the region close to the specular condition $\beta = 0$ is zero because of the beam stop which prevents the detector from being saturated.

speckle patterns with exact calculations of the scattering. We shall come back to this point at the beginning of the next section.

We shall proceed first with the explicit calculation of the coherent scattering from a single surface. To keep the results as simple as possible, certain approximations are applied. The exact result for multilayers is given in Appendix A.3. According to the formalism of the previous section, only an expression for the Fresnel density $\tilde{\varrho}_{\mathrm{F}}(\boldsymbol{K})$ is needed, which may then be inserted into Eq. (7.16). As already mentioned, in scattering experiments sample-fixed coordinates $\boldsymbol{K}' = (K_{x'}, K_{y'}, K_{z'})$ are more appropriate than $\boldsymbol{K} = (K_x, K_y, K_z)$ (see Fig. 7.2). They may easily be calculated via $K_{x'} = -K_x \sin\alpha_{\mathrm{i}} + K_z \cos\alpha_{\mathrm{i}}$, $K_{y'} = K_y$, and $K_{z'} = -K_x \cos\alpha_{\mathrm{i}} - K_z \sin\alpha_{\mathrm{i}}$, where α_{i} is the mean angle between $\boldsymbol{k}_{\mathrm{i}}$ and the surface of the sample.

The Fourier transform $\tilde{\varrho}_{\infty}(\boldsymbol{K}')$ of the ideal electron density of a single interface given by the contour $z(x', y')$ is

$$\tilde{\varrho}_{\infty}(\boldsymbol{K}') = \frac{\mathrm{i}\,\varrho_0}{K_{z'}} \iint \exp\Big\{ -\mathrm{i}\Big[K_{x'}x' + K_{y'}y' + K_{z'}\, z(x', y')\Big]\Big\}\, \mathrm{d}x'\mathrm{d}y' \,, \tag{7.19}$$

where absorption has already taken account of the finite size in the z' direction and ϱ_0 is the bulk density of the material below the surface. In the case of in-plane scattering, the directions $\boldsymbol{k}_{\mathrm{i}}$ and $\boldsymbol{k}_{\mathrm{f}}$ are given in (x, y, z) coordinates by $\boldsymbol{k}_{\mathrm{i}} = k(0, 0, 1)$ and $\boldsymbol{k}_{\mathrm{f}} = k(\sin\Phi, 0, \cos\Phi)$, where Φ denotes the angle between the mean directions of the incident and scattered radiation.

As the surface contour, an arrangement of randomly distributed steps of width x_j and heights h_j is considered. The Fourier transform of the Fresnel electron density which is required for the calculation of the scattering via Eq. (7.16) is given by (see Appendix A.3 for further details)

$$\tilde{\varrho}_{\mathrm{F}}(\boldsymbol{K}') = \frac{C_0}{2\,K_{z',x'}} \exp\left(-\frac{\mathrm{i}}{2\Omega}\frac{K_{x'}^2}{k_{\mathrm{s}}^2 + \mathrm{i}k_{L_1}^2\varepsilon^2}\right) \sum_j \exp(-\mathrm{i}\,K_{z',x'}h_j)$$
$$\times \Big\{ \Big[C(y_{j+1}) - C(y_j)\Big] + \mathrm{i}\Big[S(y_{j+1}) - S(y_j)\Big]\Big\} \,, \tag{7.20}$$

where $\varepsilon^2 = 2\pi^2/(\Omega l^2 k_{L_1}^2)$ (l is the sample size), C_0 is a constant, and $K_{z',x'}$, y_j, and y_{j+1} are given by

$$K_{z',x'} = K_{z'} - \frac{k_{\mathrm{sc}}^2}{k_{\mathrm{s}}^2 + \mathrm{i}k_{L_1}^2\varepsilon^2} K_{x'} \tag{7.21}$$

and

$$y_{j/j+1} = \frac{K_{x'} - 2\Omega k_{\mathrm{s}}^2\, x_{j/j+1} - 2\Omega k_{\mathrm{sc}}^2\, h_j}{\sqrt{2\Omega\pi k_{\mathrm{s}}^2}} \,. \tag{7.22}$$

The functions $C(y)$ and $S(y)$ are the Fresnel integrals [4, 147]

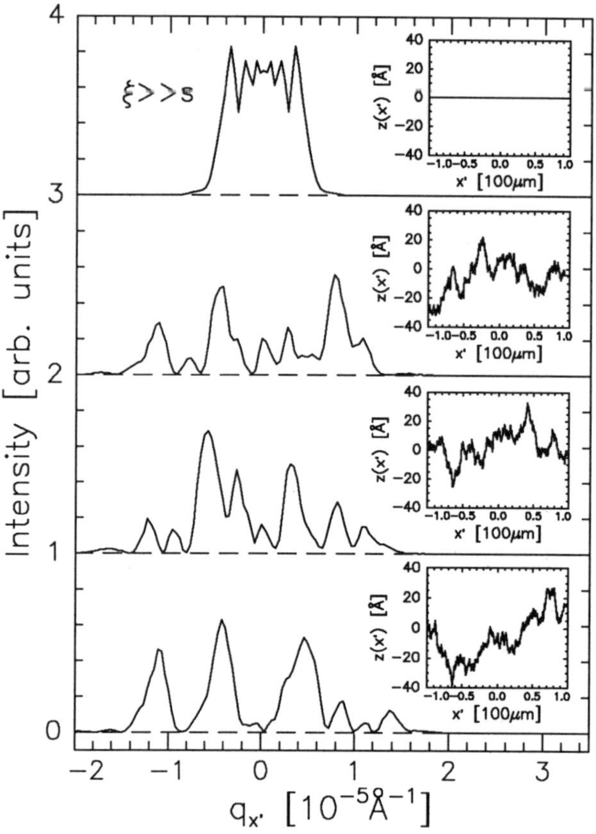

Fig. 7.7. Calculated speckle patterns for rough surfaces (coherent case, $\xi \gg s$). Transverse $q_{x'}$ scans are depicted for $L = 0.5\,\mathrm{m}$, for the surfaces shown in the respective insets. The *topmost curve* is calculated for a smooth surface. It should be noted that the height–height correlation functions $C(X') = \langle z(x')\,z(x' + X')\rangle$ of all other interfaces are identical. The other parameters are the same as used for the calculations in Fig. 7.3 [375]

$$C(y) = \frac{2}{\sqrt{\pi}} \int_0^y \cos(t^2)\,\mathrm{d}t \quad \text{and} \quad S(y) = \frac{2}{\sqrt{\pi}} \int_0^y \sin(t^2)\,\mathrm{d}t\,, \tag{7.23}$$

$k_\mathrm{s}^2 = k_{L_1}^2 \sin^2\alpha_\mathrm{i} + k_{L_2}^2 \sin^2\alpha_\mathrm{f}$, and $k_\mathrm{sc}^2 = k_{L_1}^2 \sin\alpha_\mathrm{i}\cos\alpha_\mathrm{i} - k_{L_2}^2 \sin\alpha_\mathrm{f}\cos\alpha_\mathrm{f}$. An integration over $K_{y'}$ was also performed to obtain Eq. (7.20) [375].

Figure 7.7 shows calculated $q_{x'}$ scans ($L_1 = L_2 = 0.5\,\mathrm{m}$) for various surfaces, given by the height functions $z(x')$ in the respective insets. The topmost curve is that for a smooth interface. The same parameters as for the calculations of the curves in Fig. 7.3 were used. A random arrangement of steps with widths x_j in the interval $0\,\text{Å} < x_j < 5000\,\text{Å}$ and heights h_j fulfilling the condition $|h_{j+1} - h_j| \leq 5\,\text{Å}$ was taken as the height function $z(x')$. Thus, the statistical properties of the surfaces shown in the three lower insets of Fig. 7.7 are the same. Hence they would yield the same scattering pattern if the impinging radiation was incoherent.

The curves in Fig. 7.7 were calculated for coherent surface scattering ($\xi \gg s$). They indicate that different surface morphologies which have identical statistical properties yield quite different speckle patterns.

This example demonstrates that a precise calculation of speckle patterns is very difficult. Even worse, if data are compared quantitatively with a measurement the detector resolution and the monochromaticity of the impinging radiation have to be taken into account according to Eq. (7.12). Although the first attempts can be found in the literature [384], it seems unlikely that a full reconstruction of $z(x',y')$ from a speckle pattern can be achieved. The reasons for this will be given in the next section and the main future applications of coherent x-ray scattering will be discussed.

7.4 Future Developments

The previous section has shown that the exact calculation of speckle patterns is possible[17]. However, since a surface is illuminated with a coherent x-ray beam of $\lambda \sim 1$ Å such length scales contribute to the scattering. This means that for a detailed reconstruction of the surface contour $z(x',y')$ *all* elements involved in the scattering must be known very precisely. This is, in particular, true for the apertures A and B in Fig. 7.2. Recently *Vlieg et al.* [388] demonstrated a way to control a micron-sized pinhole with high reproducibility by using a particular slit setting (see also Fig. 7.1).

Apart from the experimental challenges, which may be solved in the near future[18], another problem arises if one tries to reconstruct the exact $z(x',y')$ from data like those shown in Fig. 7.5: Since a thin oxide layer is present on silicon surfaces, the inner Si/SiO_2 interface also contributes to the scattering. Thus, the full formalism outlined in Appendix A.3 has to be used for the data analysis. This demonstrates that the advantage of having a coherent beam is a disadvantage at the same time, because everything, e.g. organic overlayers or water films, built-in sample imperfections, etc., contributes to the scattering in a rather complex manner[19].

The above arguments do not mean that coherent x-ray scattering is useless. On the contrary, the aim was to emphasize that the analysis has to be done very carefully, and that – although in principle possible – a real-space reconstruction of interface contours on a lateral angstrom scale still seems to be unlikely. But there is no question that in future these experiments may provide new and unique information about samples that cannot be obtained

17 The resultant formulas obtained without approximations are given in Appendix A.3.

18 Nowadays, with a micron-sized pinhole it is possible to produce a coherent beam with a flux of 10^9 photons/sec at the ESRF [2, 149].

19 The situation is much "better" in the case of incoherent scattering since the scattering from rough low-density layers or particles is effectively damped by the roughness factor $\exp(-q_z^2\sigma^2)$ (see Sect. 2.3).

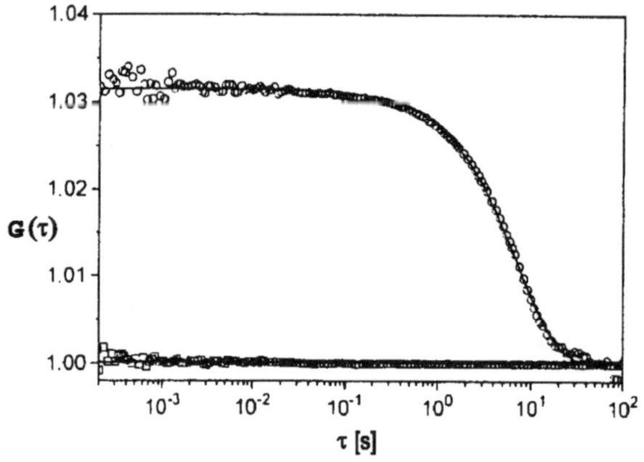

Fig. 7.8. Correlation function $\mathcal{G}(\tau)/\mathcal{G}(0)$ measured for a palladium colloid in glycerol at a wavevector $q = 1.58 \times 10^{-3}\,\text{Å}^{-1}$. *Symbols*: data. *Line*: fit with an exponential decay. *Horizontal line* at the bottom: correlation for the incident beam (figure from *Thurn-Albrecht et al.* [364])

otherwise. Particular interface structures on larger length scales (e.g. steps on miscut surfaces) may be solved by coherent x-ray-scattering experiments. However, in future the real domain of coherent x-ray scattering will be the field of spectroscopy. Here, examples of XPCS can already be found in the literature [91, 245, 364]; Fig. 7.8 depicts one of these.

With XPCS low-frequency dynamics (10^6 Hz to 10^{-3} Hz) in a q range up to several Å^{-1} may be investigated. This "window" is accessible neither by neutron scattering, which allows the same q range but is sensitive to higher frequencies, nor by light scattering, which covers low-frequency dynamics but only for the long-wavelength range $q < 10^{-3}\,\text{Å}^{-1}$ in materials that do not absorb visible light [149].

Figure 7.8 shows a normalized time correlation function $\mathcal{G}(\tau)/\mathcal{G}(0) = \langle I(t)\, I(t+\tau)\rangle / \langle I^2(t)\rangle$ measured by XPCS at a wavevector transfer $q = 1.58 \times 10^{-3}\,\text{Å}^{-1}$ for a palladium colloid in glycerol (*Thurn-Albrecht et al.* [364]). For a diffusion process the expected functional form of the correlation function is [35]

$$\mathcal{G}(\tau)/\mathcal{G}(0) = A(q)\, \exp(-2q^2 D\tau) + 1\,, \tag{7.24}$$

where D is the diffusion constant for the process under investigation.

A scattering theory for XPCS going beyond Eq. (7.24) and based on the results of the previous section will now be briefly outlined. As in the case of optical photon correlation spectroscopy [64], XPCS probes the dynamic properties of matter by analyzing the temporal correlations among scattered photons. Mathematically, correlations of the scattered intensity are described by the time correlation function $\langle I(t)\, I(t+\tau)\rangle$, where $\langle \ldots \rangle$ denotes a suitably long time average over times much longer than τ, and $I(t)$ represents the intensity measured in the detector at time t. In contrast to the theory outlined in the previous sections, the electron density $\varrho(\boldsymbol{r}, t)$ is now a time-dependent quantity.

It is further assumed that time averages can be performed independently over the sample electron density fluctuations and the incident-beam fluctuations, since these are assumed to be independent. It is also assumed that τ is large enough that time correlations in the incident-beam intensity have decayed on the timescale measured in the experiment[20]. A similar calculation to that in Sect. 7.2 yields [377]

$$\langle I(t)\, I(t+\tau)\rangle = \frac{r_e^4 \bar{\omega}^2}{4\pi^2} \frac{I^2/A^2}{(\lambda L_1 L_2)^4} \mathcal{G}(\boldsymbol{q},\tau) * R_{\mathrm{B}} * R_{\mathrm{B}'} \,, \tag{7.25}$$

where $*R_{\mathrm{B}} * R_{\mathrm{B}'}$ implies a double folding with the detector aperture resolution function[21] for a given nominal wavevector transfer $\boldsymbol{q}$ according to Eq. (7.12), and

$$\begin{aligned}\mathcal{G}(\boldsymbol{K},\tau) = & \iint \tilde{F}[\bar{\omega}(1-\Omega)]\tilde{F}[\bar{\omega}(1-\Omega')] \\ & \times\Bigg(\iiiint \Psi(\boldsymbol{s})\Psi^*(\boldsymbol{s}')g(\boldsymbol{s}-\boldsymbol{s}')\Psi(\boldsymbol{s}_1)\Psi^*(\boldsymbol{s}_1')g(\boldsymbol{s}_1-\boldsymbol{s}_1') \\ & \times \exp\left\{\mathrm{i}\frac{\Omega}{2}k_{L_1}^2\left(s^2-s'^2\right)\right\}\exp\left\{\mathrm{i}\frac{\Omega'}{2}k_{L_1}^2\left(s_1^2-s_1'^2\right)\right\} \\ & \times\Big\langle \tilde{\varrho}_{\mathrm{F}}(\boldsymbol{K}+\Omega k_{L_1}^2\boldsymbol{s},t) \\ & \times \tilde{\varrho}_{\mathrm{F}}^*(\boldsymbol{K}+\Omega k_{L_1}^2\boldsymbol{s}',t)\,\tilde{\varrho}_{\mathrm{F}}(\boldsymbol{K}+\Omega' k_{L_1}^2\boldsymbol{s}_1,t+\tau) \\ & \times \tilde{\varrho}_{\mathrm{F}}^*(\boldsymbol{K}+\Omega' k_{L_1}^2\boldsymbol{s}_1',t+\tau)\Big\rangle \mathrm{d}\boldsymbol{s}\mathrm{d}\boldsymbol{s}'\mathrm{d}\boldsymbol{s}_1\mathrm{d}\boldsymbol{s}_1'\Bigg)\,\mathrm{d}\Omega\mathrm{d}\Omega' \,,\end{aligned} \tag{7.26}$$

where the notation of Sect. 7.2 is used. In Eq. (7.26) the time dependence of the Fourier coefficients of $\tilde{\varrho}_{\mathrm{F}}(\boldsymbol{K})$ has been introduced explicitly. The subscript “F” again means that the Fourier transform of the Fresnel electron density has to be taken (see Eq. 7.15). We recapitulate that the arguments of the Fourier transform of the Fresnel density in Eq. (7.26) represent the Cartesian components $\boldsymbol{K} = (K_x, K_y, K_z)$ relative to the (x, y, z) axes shown in Fig. 7.2. The function $\tilde{F}[\bar{\omega}(1-\Omega)]$ is the Fourier transform of the time autocorrelation function $F(\tau) = \exp(-\tau/\tau_1)$ of the source, i.e. it corresponds to the Lorentzian of Eq. (7.12).

In Appendix A.4 Eq. (7.26) is further simplified for N independently moving particles, finally yielding

$$\mathcal{G}(\boldsymbol{K},\tau) = N^2 A^2 + N^2 \int \tilde{F}[\bar{\omega}(1-\Omega)]$$

[20] Incident-beam intensity fluctuations are, in general, much more rapid than the sample dynamics which are to be probed.

[21] The double folding is neglected in the following.

$$\times\left(\iint \exp\left\{\mathrm{i}\Omega k_{L_1}^2(s^2-s'^2)\right\}\left[\Psi(\boldsymbol{s})\right]^2\left[\Psi^*(\boldsymbol{s}')\right]^2\left[g(\boldsymbol{s}-\boldsymbol{s}')\right]^2\right.$$

$$\left.\times\left[\Lambda(\boldsymbol{K}+\Omega k_{L_1}^2\boldsymbol{s},\boldsymbol{K}+\Omega k_{L_1}^2\boldsymbol{s}',\tau)\right]^2\mathrm{d}\boldsymbol{s}\mathrm{d}\boldsymbol{s}'\right)\mathrm{d}\Omega\,, \qquad (7.27)$$

where $\Lambda(\boldsymbol{K}_\alpha, \boldsymbol{K}_\beta, \tau)$ is given by Eq. (A.29) in Appendix A.4. The second term in Eq. (7.27) yields the time dependence of $\mathcal{G}(\boldsymbol{K}, \tau)$. Therefore Eq. (7.27) describes a time-dependent signal riding on a time-independent "background". Hence, Eq. (7.27) is an analytical expression for the time correlation function $\mathcal{G}(\boldsymbol{K}, \tau)$, where the coherence factor $g(\boldsymbol{s} - \boldsymbol{s}')$, the aperture function $\Psi(\boldsymbol{s})$, and the longitudinal coherence length $\xi_l = \tau_l c$ have to be inserted. These quantities are determined by the particular experimental conditions. With Eq. (7.25) and Eq. (7.26) or Eq. (7.27), a rather general and quantitative treatment of XPCS data will become possible in future.

8. Closing Remarks

Surface- and interface-sensitive x-ray-scattering techniques play an important role in many areas of physics. This book was intended to give a review of their application to problems concerning soft-matter thin films. Many fundamental questions such as the change of the wavenumber spectrum of capillary waves due to a background potential and the altering of the packing density of a fluid near a hard wall have been discussed. Also, problems that arise in materials science, such as the question how strongly roughness is vertically propagated in thin films and the detection of disorder in LB multilayers, have been in the focus of this work. Research will certainly continue in these fields with the "conventional" x-ray reflectivity and diffuse-scattering methods described in Chaps. 2–6.

X-ray scattering using coherent beams, as presented in Chap. 7, is a promising new field for the near future – this is, in particular, true for the investigation of soft-matter thin films. Although we are still far away from the flux of coherent photons that a common laser emits in the range of visible light, the experiments that have already been carried out within the last years are very impressive. At all operating synchrotron facilities worldwide, beamlines are under construction that will be dedicated to experiments with coherent x-rays.

A promising new tool that may be used in the near future to solve interface structures is multiple-energy x-ray holography. This technique, developed recently by *Gog et al.* [143], allows a three-dimensional imaging of structures with atomic resolution. Here many possible applications in soft-matter physics are already under consideration.

Last but not least, the question of industrial applications will be addressed. Reflectivity is already in use to characterize the interfaces of thin films. Complete "reflectivity setups" consisting of an x-ray generator, a diffractometer, detectors, and the analyzing software are available. However, it still needs quite a bit of experience to extract reliable information from reflectivity data that goes beyond the simple determination of a layer thickness. Here data-analyzing techniques such those outlined in Chap. 4 may improve this situation in the future.

A. Appendix

A.1 The Hilbert Phase of Reflection Coefficients

In this appendix we discuss under what conditions the phase of the structure factor $F(q_z)$ (see Sect. 4.1 and Eq. 4.1) may be calculated unambiguously from a reflectivity measurement, i.e. from the modulus of $F(q_z)$. Here the argument given by *Clinton* [60] is basically followed.

At the beginning of the considerations stands the Titchmarsh theorem: If a function $f(z)$ is zero on some domain, say $f(z) = 0$ for $z < 0$, then its Fourier transform

$$g(q) = \int_0^\infty f(z)\exp(\mathrm{i}\,qz)\,\mathrm{d}z \tag{A.1}$$

is analytic in the upper half of the complex q plane. From the Cauchy principal-value theorem it follows that $\mathrm{Re}\{g(q)\}$ and $\mathrm{Im}\{g(q)\}$ are Hilbert transforms. The Kramers–Kronig relation, which connects $f'_j(E)$ and $f''_j(E)$ in Eqs. (2.4) and (2.5) in Sect. 2.1, for example, is one of the most prominent application of the Titchmarsh theorem in solid-state physics.

Since a substrate is always present when scattering from thin films is considered, one has $\varrho(z) = \varrho_{\mathrm{subst.}} = \mathrm{const.}$ for large negative z values, thus $\mathrm{d}\varrho(z)/\mathrm{d}z = 0$, and if the coordinate system is defined in an appropriate manner this holds for all $z < 0$. Then the Titchmarsh theorem ensures that $F(q_z)$, as given by Eq. (4.1), is analytic in the upper half-plane (UHP)[1] and, moreover, that $\mathrm{Re}\{F(q_z)\}$ and $\mathrm{Im}\{F(q_z)\}$ are Hilbert transforms, which are equivalent to dispersion relations. Furthermore, $\ln[F(q_z)]$ is analytic, too, except at those values of q_z in the UHP where $F(q_z) = 0$. If $F(q_z) \neq 0$ for all q_z with $\mathrm{Im}\{q_z\} > 0$, then the so-called Hilbert phase

$$\phi_{\mathrm{H}}(q_z) = -\frac{2\,q_z}{\pi}\int_0^\infty \frac{\ln[|F(q'_z)|/|F(q_z)|]}{q'^2_z - q^2_z}\,\mathrm{d}q'_z \tag{A.2}$$

and the exact phase of $F(q_z) = |F(q_z)|\exp\{i\phi(q_z)\}$

[1] It should be emphasized here that q_z has to be treated as a complex variable. This is not a requirement due to absorption or evanescent waves.

$$\phi(q_z) = \arctan\left(\frac{\mathrm{Im}\{F(q_z)\}}{\mathrm{Re}\{F(q_z)\}}\right) \tag{A.3}$$

agree. This remarkable result transforms the phase problem to the condition of whether or not $F(q_z)$ has zeros in the UHP. In the latter case the x-ray reflectivity given by Eq. (4.1) in Sect. 4.1 is determined by $|F(q_z)|$ alone, from which the phase may be calculated via Eq. (A.2). A simple way to perform a numerical calculation of the Hilbert phase $\phi_{\mathrm{H}}(q_z)$ avoiding the singularity in the denominator of the integral is given by *Reiss* [296].

Hence, profiles have to be found which fulfill the condition $F(q_z) \neq 0$ in the UHP. Here *Clinton* [60] has proven that a very interesting class of functions obeys this requirement. For a layer stack with N sharp interfaces and constant electron densities ϱ_m, the derivative of $\varrho(z)$ is

$$\frac{\mathrm{d}\varrho(z)}{\mathrm{d}z} = \sum_{m=1}^{N} \Delta\varrho_m \, \delta(z - z_m) \,, \tag{A.4}$$

where the density contrasts are $\Delta\varrho_m = \varrho_{m+1} - \varrho_m$ and the interface locations are z_m. This profile yields a reflectivity from which the phase may be reconstructed unambiguously via Eq. (A.2) if the density contrast at the substrate $\Delta\varrho_1$ fulfills the condition

$$|\Delta\varrho_1| > \sum_{m=2}^{N} |\Delta\varrho_m| \,. \tag{A.5}$$

All arguments given above are also valid if the exact dynamical treatment of the reflectivity as presented in Chap. 2 is performed. However, since this complicates the calculations considerably we will not go into more details here. The reader is referred to the book by *Chadan & Sabatier* [56] for the general formalism and to the work of *Lipperheide et al.* [219, 220] for examples concerning x-ray and neutron reflectivity. *Klibanov & Sacks* [192] have published a very sophisticated algorithm that allows one to reconstruct a scattering potential from a reflectivity measurement when a small fraction of this potential is known beforehand. Also, the influence of zeros of $F(q_z)$ in the UHP is discussed and a very promising new approach for incorporating such zeros into an inversion algorithm is given in Ref. [99].

A.2 The Formalism of the DWBA

The basic mathematics of the DWBA in the case of layer systems including vertical roughness correlations is given in this appendix. Details of the calculations may be found in the papers of *Sinha et al.* [335] and *Holý et al.* [164, 165]. A homogeneous N-layer system with mean positions z_j, refractive indices n_j, and perfectly smooth interfaces is used as an unperturbed ideal system. The Fresnel theory (*Parratt* formalism, Sect. 2.2) yields the corresponding eigenfunctions $|\psi_\mathrm{i}\rangle$ and $|\psi_\mathrm{f}\rangle$. This system is now "perturbed" by the roughnesses σ_j of the layers (for the notation see Sect. 2.2). Owing to the roughness, there is a nonvanishing probability of a transition between two eigenstates, and off-specular scattering with $\alpha_\mathrm{i} \neq \alpha_\mathrm{f}$ becomes possible.

The starting point of the calculations is the Helmholtz equation which is now written in the following form:

$$(\nabla^2 + k^2)|\,\psi\rangle = V(\boldsymbol{r})\,|\psi\rangle\,, \tag{A.6}$$

where the scattering potential is $V(\boldsymbol{r}) = k^2[1 - n^2(\boldsymbol{r})]$. The potential may be split into a part $\bar{V}(\boldsymbol{r})$ corresponding to the sharp-interface system, and the (small) perturbation $\delta V(\boldsymbol{r})$ caused by the roughnesses.

The transition matrix element $\langle \mathrm{f}\,|T|\,\mathrm{i}\rangle$ for a scattering process from the initial state $|\mathrm{i}\rangle$ into the final state $|\mathrm{f}\rangle$ is given by [242]

$$\langle \mathrm{f}\,|T|\,\mathrm{i}\rangle = \langle\widetilde{\psi}_\mathrm{f}|\bar{V}|\phi_\mathrm{i}\rangle + \langle\widetilde{\psi}_\mathrm{f}|\delta V|\chi\rangle \tag{A.7}$$

$$\approx \underbrace{\langle\widetilde{\psi}_\mathrm{f}|\bar{V}|\phi_\mathrm{i}\rangle}_{\bar{V}_\mathrm{if}} + \underbrace{\langle\widetilde{\psi}_\mathrm{f}|\delta V|\psi_\mathrm{i}\rangle}_{\delta V_\mathrm{if}}\,, \tag{A.8}$$

where

$$\phi_\mathrm{i}(\boldsymbol{r}) = \exp(\mathrm{i}\,\boldsymbol{k}_\mathrm{i}\cdot\boldsymbol{r})\,, \tag{A.9}$$

$$\psi_\mathrm{i}(\boldsymbol{r}) = T_\mathrm{i}(z)\,\exp\{\mathrm{i}\,\boldsymbol{k}_\mathrm{i}(z)\cdot\boldsymbol{r}\} + R_\mathrm{i}(z)\,\exp\{\mathrm{i}\,\boldsymbol{k}_\mathrm{i}'(z)\cdot\boldsymbol{r}\}\,, \tag{A.10}$$

$$\widetilde{\psi}_\mathrm{f}(\boldsymbol{r}) = T_\mathrm{f}^*(z)\,\exp\{\mathrm{i}\,\boldsymbol{k}_\mathrm{f}^*(z)\cdot\boldsymbol{r}\} + R_\mathrm{f}^*(z)\,\exp\{\mathrm{i}\,\boldsymbol{k}_\mathrm{f}'^*(z)\cdot\boldsymbol{r}\}\,. \tag{A.11}$$

The impinging plane wave is given by $|\phi_\mathrm{i}\rangle$, and $|\psi_\mathrm{i}\rangle$ and $|\widetilde{\psi}_\mathrm{f}\rangle$ are two eigenstates of the ideal system, where the latter is time-reversed. Equation (A.8) is the central approximation of the DWBA: The exact, but unknown eigenstate $|\chi\rangle$ of the system including roughness is replaced by the state $|\psi_\mathrm{i}\rangle$. The amplitudes $R_\mathrm{i}, R_\mathrm{f}$ and $T_\mathrm{i}, T_\mathrm{f}$ are calculated from the equations of the *Parratt* formalism for layer systems (see Sect. 2.2, Eqs. 2.16–2.20). The intensity scattered into a solid angle $\mathrm{d}\Omega$ is proportional to the cross section $\mathrm{d}\sigma/\mathrm{d}\Omega$ and may be calculated from "Fermi's golden rule"[2]

$$\frac{\mathrm{d}\sigma}{\mathrm{d}\Omega} = \frac{1}{16\pi^2}\left\langle |\langle \mathrm{f}|T|\mathrm{i}\rangle|^2\right\rangle = \frac{1}{16\pi^2}\left\langle \left|\bar{V}_\mathrm{if} + \delta V_\mathrm{if}\right|^2\right\rangle\,, \tag{A.12}$$

where Eq. (A.8) has been used. This cross section can be separated into a specular and a diffuse part,

$$\left(\frac{\mathrm{d}\sigma}{\mathrm{d}\Omega}\right)_\mathrm{spec.} = \frac{\left|\bar{V}_\mathrm{if} + \langle\delta V_\mathrm{if}\rangle\right|^2}{16\pi^2}\,, \tag{A.13}$$

$$\left(\frac{\mathrm{d}\sigma}{\mathrm{d}\Omega}\right)_\mathrm{diff.} = \frac{\left\langle\left|\delta V_\mathrm{if}\right|^2\right\rangle - \left|\left\langle\delta V_\mathrm{if}\right\rangle\right|^2}{16\pi^2}\,. \tag{A.14}$$

[2] The large brackets denote an ensemble average.

Table A.1. Two possible analytic continuations and the respective wavevector transfers. They fulfill the reciprocity theorem of optics.

$G_j^0 = T_{\mathrm{i};j+1}T_{\mathrm{f};j+1}$	$\boldsymbol{q}_j^0 = \boldsymbol{k}_{\mathrm{i};j+1} + \boldsymbol{k}_{\mathrm{f};j+1}$	$G_j^0 = T_{\mathrm{i};j}T_{\mathrm{f};j}$	$\boldsymbol{q}_j^0 = \boldsymbol{k}_{\mathrm{i};j} + \boldsymbol{k}_{\mathrm{f};j}$
$G_j^1 = T_{\mathrm{i};j+1}R_{\mathrm{f};j+1}$	$\boldsymbol{q}_j^1 = \boldsymbol{k}_{\mathrm{i};j+1} - \boldsymbol{k}_{\mathrm{f};j+1}$	$G_j^1 = T_{\mathrm{i};j}R_{\mathrm{f};j}$	$\boldsymbol{q}_j^1 = \boldsymbol{k}_{\mathrm{i};j} - \boldsymbol{k}_{\mathrm{f};j}$
$G_j^2 = R_{\mathrm{i};j+1}T_{\mathrm{f};j+1}$	$\boldsymbol{q}_j^2 = -\boldsymbol{q}_j^1$	$G_j^2 = R_{\mathrm{i};j}T_{\mathrm{f};j}$	$\boldsymbol{q}_j^2 = -\boldsymbol{q}_j^1$
$G_j^3 = R_{\mathrm{i};j+1}R_{\mathrm{f};j+1}$	$\boldsymbol{q}_j^3 = -\boldsymbol{q}_j^0$	$G_j^3 = R_{\mathrm{i};j}R_{\mathrm{f};j}$	$\boldsymbol{q}_j^3 = -\boldsymbol{q}_j^0$

The specular part is given by Eq. (A.13), and Eq. (A.14) yields the cross section for diffuse scattering as given by Eqs. (6.11) and (6.12) in Sect. 6.1.2. The indices j and k in Eqs. (6.11) and (6.12) in the perpendicular wavevector transfers mean that these quantities have to be evaluated in the respective layer. The upper index m or n indicates one of the combinations given in Table A.1, from which also the the quantities $\widetilde{G}_j^m$ in Eqs. (6.11) and (6.12), together with the definition $\widetilde{G}_j^m = G_j^m \exp(-\mathrm{i}\, q_{z,j}^m z_j)$, may be obtained. The functions in Table A.1 are analytic continuations of the wave functions at the respective interfaces [164, 165]. The deviation of the DWBA result from the exact unknown solution depends on the eigenfunctions given by Eqs. (A.9)–(A.11), which were chosen to calculate the transition matrix elements. Since a sharp-interface system is used for the unperturbed potential, it is expected that these deviations increase with increasing $q_z\sigma$. While this is true for the reflectivity [335], which may be also calculated within the DWBA, surprisingly this is not the case for the diffuse-scattering cross section: In the limit of large q_z it reduces to the correct kinematical result.

A.3 Exact Formulas for Coherent X-Ray Scattering

The Fourier transform of the Fresnel electron density for a system consisting of N homogeneous layers with average densities ϱ_j, mean interface locations μ_j, and interface contour functions $z_j(x', y')$ will be calculated here[3]. The Fourier transform $\tilde{\varrho}_\infty(\boldsymbol{K}')$ of the ideal electron density, i.e. that for an infinitely large sample, of such a system is given by

$$\tilde{\varrho}_\infty(\boldsymbol{K}') = \frac{\mathrm{i}}{K_{z'}} \sum_{j=1}^{N} \Delta\varrho_j \exp(-iK_{z'}\mu_j) \tag{A.15}$$
$$\times \iint \exp\Big\{ -\mathrm{i} \Big[K_{x'}x' + K_{y'}y' + K_{z'}\, z_j(x', y')\Big]\Big\} \,\mathrm{d}x'\mathrm{d}y' ,$$

where $\Delta\varrho_j = \varrho_{j+1} - \varrho_j$ is the electron density contrast between the layers. Here it should be noted that the description given by Eq. (A.15) already

[3] The $z_j(x', y')$ are assumed to have zero mean.

includes finite-size effects in the z' direction (absorption). The Fourier transform of the Fresnel density may be calculated via two convolutions (indicated by $*$) [340]:

$$\tilde{\varrho}_{\mathrm{F}}(\boldsymbol{K}') = \tilde{\varrho}_{\infty}(\boldsymbol{K}') * \tilde{\mathcal{T}}(\boldsymbol{K}') * \tilde{\mathcal{F}}(\boldsymbol{K}') \,, \tag{A.16}$$

where $\tilde{\mathcal{F}}(\boldsymbol{K}')$ is the Fourier transform of the phase factor in Eq. (7.15) and $\tilde{\mathcal{T}}(\boldsymbol{K}')$ accounts for the finite size of the sample in the x' and y' directions given by the Fourier transform of the real-space truncation function $\mathcal{T}(\boldsymbol{r}') = \exp\{-(x'^2 + y'^2)\pi^2/l^2\}$, where l is the extent of the sample.

For reasons of simplicity, only the case of in-plane scattering will be discussed. In this case the directions $\boldsymbol{k}_{\mathrm{i}}$ and $\boldsymbol{k}_{\mathrm{f}}$ are given in (x, y, z) coordinates by $\boldsymbol{k}_{\mathrm{i}} = k(0,0,1)$ and $\boldsymbol{k}_{\mathrm{f}} = k(\sin\Phi, 0, \cos\Phi)$, where Φ denotes the angle between the mean directions of the incident and scattered radiation (often also called 2θ).

After some algebra one obtains the rather complicated result [375]

$$\begin{aligned}\tilde{\varrho}_{\mathrm{F}}(\boldsymbol{K}') = \tilde{\varrho}_{(000)}(\boldsymbol{K}') + C_0 \iint \exp\left\{ - \frac{\mathrm{i}}{2\Omega} \frac{\tilde{K}_{x'}^2}{k_{\mathrm{s}}^2 + \mathrm{i}k_{L_1}^2 \varepsilon^2} \right\} \\ \times \exp\left\{ - \frac{\mathrm{i}\,\tilde{K}_{y'}^2}{2\Omega(k_{L_1}^2 + k_{L_2}^2) + \mathrm{i}4\pi^2/l^2} \right\} \\ \times \sum_{j=1}^{N} \Delta\varrho_j \tilde{H}_j \left\{ \tilde{K}_{x'}, \tilde{K}_{y'}, \boldsymbol{K}'; [z_j] \right\} \mathrm{d}\tilde{K}_{x'}\, \mathrm{d}\tilde{K}_{y'} \,,\end{aligned} \tag{A.17}$$

where $\varepsilon^2 = 2\pi^2/(\Omega l^2 \, k_{L_1}^2)$ and the pre-factor C_0 is given by

$$C_0 = \frac{(2\pi)^{-2}(2\Omega)^{-1/2}\, l^3}{[2\Omega l^2(k_{L_1}^2 + k_{L_2}^2) + 4\pi^2 \mathrm{i}]^{1/2}(k_{\mathrm{s}}^2 + \mathrm{i}k_{L_1}^2 \varepsilon^2)^{1/2}} \,. \tag{A.18}$$

The interface contours $z_j(x', y')$ enter into the result via the function

$$\begin{aligned}\tilde{H}_j\Big\{\ldots\Big\} = \iint \exp\Big\{ - \mathrm{i}\,(\tilde{K}_{x'}x' + \tilde{K}_{y'}y') + K_{x'}x' + K_{y'}y') \Big\} \\ \times \frac{\exp\Big\{ - \mathrm{i} \Big[K_{z',\tilde{x}'} + \eta_j(x', y')/2 \Big] \Big[z_j(x', y') + \mu_j \Big] \Big\}}{K_{z',\tilde{x}'} - \eta_j(x', y')} \mathrm{d}x' \mathrm{d}y' \,,\end{aligned} \tag{A.19}$$

where the abbrevations

$$K_{z',\tilde{x}'} = K_{z'} + \frac{k_{\mathrm{sc}}^2}{k_{\mathrm{s}}^2 + \mathrm{i}k_{L_1}^2 \varepsilon^2} \tilde{K}_{x'} \,, \tag{A.20}$$

$$\eta_j(x', y') = 2\Omega k_{L_1}^2 \frac{k_{L_2}^2 \sin^2\Phi + \mathrm{i}k_{\mathrm{c}}^2 \varepsilon^2}{k_{\mathrm{s}}^2 + \mathrm{i}k_{L_1}^2 \varepsilon^2} \Big[z_j(x', y') + \mu_j \Big] \,, \tag{A.21}$$

and

$$k_{\mathrm{c}}^2 = k_{L_1}^2 \cos^2\alpha_{\mathrm{i}} + k_{L_2}^2 \cos^2\alpha_{\mathrm{f}} \;, \tag{A.22}$$
$$k_{\mathrm{s}}^2 = k_{L_1}^2 \sin^2\alpha_{\mathrm{i}} + k_{L_2}^2 \sin^2\alpha_{\mathrm{f}} \;, \tag{A.23}$$
$$k_{\mathrm{sc}}^2 = k_{L_1}^2 \sin\alpha_{\mathrm{i}}\cos\alpha_{\mathrm{i}} - k_{L_2}^2 \sin\alpha_{\mathrm{f}}\cos\alpha_{\mathrm{f}} \;, \tag{A.24}$$

have been used. In Eq. (A.17), $\tilde{\varrho}_{(000)}(\boldsymbol{K}')$ describes the Bragg–Fresnel scattering around the origin of reciprocal space and does not need to be considered any further [340]. The surface and interface part is given by the second part, i.e. the integrals over $\tilde{K}_{x'}$ and $\tilde{K}_{y'}$ and the sum over the individual interface contributions. Since $|k_{L_{1,2}}^2[z_j(x',y') + \mu_j]| \ll |K_{z',\tilde{x}'}|$ is often well satisfied, the quantity $\eta_j(x',y')$ may be neglected. The simplifies the resultant formula slightly. Equation (A.17) is the exact expression for the Fourier transform of the Fresnel electron density of a multilayer which is needed for the precise calculation of speckle patterns via Eqs. (7.12) and (7.16) in Sect. 7.2.

A remarkable fact which follows from Eqs. (A.17)–(A.21) is that the combination $K_{z',\tilde{x}'}$ of $K_{z'}$ and $K_{x'}$ (see Eq. A.20) enters directly into the result, making it clear that there is no separation between "specular" and "diffuse" scattering.

A.4 Diffusive Particle Motion and XPCS

The result given by Eq. (7.26) in Sect. 7.4 is rather complex. A way to simplify this equation will be sketched in this appendix. The time average in Eq. (7.26) may be evaluated by using the decoupling approximation

$$\begin{aligned}
\Big\langle \dots \Big\rangle &= \Big\langle \tilde{\varrho}_{\mathrm{F}}(\boldsymbol{K}_1,t)\,\tilde{\varrho}_{\mathrm{F}}^*(\boldsymbol{K}_2,t)\,\tilde{\varrho}_{\mathrm{F}}(\boldsymbol{K}_3,t+\tau)\,\tilde{\varrho}_{\mathrm{F}}^*(\boldsymbol{K}_4,t+\tau)\Big\rangle \qquad \text{(A.25)}\\
&\approx \Big\langle \tilde{\varrho}_{\mathrm{F}}(\boldsymbol{K}_1,t)\tilde{\varrho}_{\mathrm{F}}^*(\boldsymbol{K}_2,t)\Big\rangle\Big\langle \tilde{\varrho}_{\mathrm{F}}(\boldsymbol{K}_3,t+\tau)\tilde{\varrho}_{\mathrm{F}}^*(\boldsymbol{K}_4,t+\tau)\Big\rangle\\
&\quad +\Big\langle \tilde{\varrho}_{\mathrm{F}}(\boldsymbol{K}_1,t)\tilde{\varrho}_{\mathrm{F}}(\boldsymbol{K}_3,t+\tau)\Big\rangle\Big\langle \tilde{\varrho}_{\mathrm{F}}^*(\boldsymbol{K}_2,t)\tilde{\varrho}_{\mathrm{F}}^*(\boldsymbol{K}_4,t+\tau)\Big\rangle\\
&\quad +\Big\langle \tilde{\varrho}_{\mathrm{F}}(\boldsymbol{K}_1,t)\tilde{\varrho}_{\mathrm{F}}^*(\boldsymbol{K}_4,t+\tau)\Big\rangle\Big\langle \tilde{\varrho}_{\mathrm{F}}^*(\boldsymbol{K}_2,t)\tilde{\varrho}_{\mathrm{F}}(\boldsymbol{K}_3,t+\tau)\Big\rangle .
\end{aligned}$$

If $\tilde{\varrho}_{\mathrm{F}}(\boldsymbol{K},t)$ describes the random motion of N independent particles, Eq. (A.25) may explicitly be shown to be rigorously true as $N \to \infty$. Equation (A.25) may also be used to describe correlation functions for collective fluctuations. If the case of N independently moving particles is considered then $\varrho(\boldsymbol{r},t)$ is simply $\sum_m \delta[\boldsymbol{r} - \boldsymbol{r}_m(t)]$, where $\boldsymbol{r}_m(t)$ is the position of particle m at time t. The Fourier transform of $\varrho_{\mathrm{F}}(\boldsymbol{r},t)$ is

$$\tilde{\varrho}_{\mathrm{F}}(\boldsymbol{K},t) = \sum_{m=1}^{N} \exp\Big\{ -\mathrm{i}\,\boldsymbol{K}\cdot\boldsymbol{r}_m(t)\Big\} * \tilde{\mathcal{F}}(\boldsymbol{K}) \;, \tag{A.26}$$

where $\tilde{\mathcal{F}}(\boldsymbol{K})$ is the Fourier transform of the phase factor in Eq. (7.15) (see Sect. 7.2). For large N the following result is obtained:

$$\Big\langle \dots \Big\rangle \approx N^2 \Big[\delta(\boldsymbol{K}_1 - \boldsymbol{K}_2)\delta(\boldsymbol{K}_3 - \boldsymbol{K}_4) + \Lambda_1(\boldsymbol{K}_1, \boldsymbol{K}_3, \tau)\, \Lambda_1^*(\boldsymbol{K}_2, \boldsymbol{K}_4, \tau) + \Lambda(\boldsymbol{K}_1, \boldsymbol{K}_4, \tau)\, \Lambda^*(\boldsymbol{K}_2, \boldsymbol{K}_3, \tau)\Big] , \tag{A.27}$$

where

$$\Lambda_1(\boldsymbol{K}_\alpha, \boldsymbol{K}_\beta, \tau) = \frac{1}{2\pi^2} \int \Big\langle \exp\Big[- \mathrm{i}\, \tilde{\boldsymbol{K}} \Delta \boldsymbol{r}(\tau)\Big]\Big\rangle_N \times \tilde{\mathcal{F}}(\tilde{\boldsymbol{K}} + \boldsymbol{K}_\alpha)\tilde{\mathcal{F}}(\tilde{\boldsymbol{K}} - \boldsymbol{K}_\beta)\, \mathrm{d}\tilde{\boldsymbol{K}}, \tag{A.28}$$

$$\Lambda(\boldsymbol{K}_\alpha, \boldsymbol{K}_\beta, \tau) = \frac{1}{2\pi^2} \int \Big\langle \exp\Big[+ \mathrm{i}\, \tilde{\boldsymbol{K}} \Delta \boldsymbol{r}(\tau)\Big]\Big\rangle_N \times \tilde{\mathcal{F}}(\tilde{\boldsymbol{K}} - \boldsymbol{K}_\alpha)\tilde{\mathcal{F}}^*(\tilde{\boldsymbol{K}} - \boldsymbol{K}_\beta)\, \mathrm{d}\tilde{\boldsymbol{K}} \,. \tag{A.29}$$

Here $\langle \dots \rangle_N$ represents an average over the N particles and $\Delta \boldsymbol{r}(\tau)$ is the distance a particle moves in time τ.

If the $\boldsymbol{K}_1, \dots, \boldsymbol{K}_4$ are substituted according to Eq. (7.26) and if it is noticed that the $\tilde{\mathcal{F}}(\boldsymbol{K})$ function is sharply peaked around $\boldsymbol{K} = 0$, then one can see that both $\tilde{\mathcal{F}}$ functions in the $\Lambda_1(\boldsymbol{K}_{1,2}, \boldsymbol{K}_{3,4}, \tau)$ integral of Eq. (A.28) have essentially no overlap since $\Omega \approx \Omega'$ by virtue of the longitudinal resolution functions. Thus $|\Lambda_1(\boldsymbol{K}_{1,2}, \boldsymbol{K}_{3,4}, \tau)| \approx 0$ is a good approximation. For the same reason the main contribution of the $\Lambda(\boldsymbol{K}_{1,2}, \boldsymbol{K}_{4,3}, \tau)$ terms in Eq. (A.27) is expected when $\boldsymbol{s} = \boldsymbol{s}_1$ and $\boldsymbol{s}' = \boldsymbol{s}_1'$. Substituting in Eq. (7.26) leads finally to Eq. (7.27) for $\mathcal{G}(\boldsymbol{K}, \tau)$. For diffusive particle motion, $\Delta \boldsymbol{r}(\tau)$ is a random variable so that

$$\Big\langle \exp\{\pm \mathrm{i}\, \boldsymbol{K} \cdot \Delta \boldsymbol{r}(\tau)\}\Big\rangle_N = \exp\Big\{ - \frac{K^2}{2} \Big\langle [\Delta \boldsymbol{r}(\tau)]^2 \Big\rangle \Big\} , \tag{A.30}$$

where $\langle [\Delta \boldsymbol{r}(\tau)]^2 \rangle = D\tau$, with D being the diffusion constant. From Eq. (A.30), an analytical expression for $\Lambda(\boldsymbol{K}_\alpha, \boldsymbol{K}_\beta, \tau)$ and hence for $\mathcal{G}(\boldsymbol{K}, \tau)$ may be obtained.

References

1. F. Abelès; Ann. Physique (Paris) **5**, 596 (1950).
2. D.L. Abernathy, G. Grübel, S.G.J. Mochrie, A.R. Sandy, G.B. Stephenson, N. Mulders, M. Sutton, S. Brauer; IUCr Synchrotron Radiation Satellite Meeting, APS/ANL, Argonne (1996).
3. F.F. Abraham; J. Chem. Phys. **68**, 3713 (1978).
4. M. Abramowitz, I.A. Stegun (eds.); *Handbook of Mathematical Functions*, Dover Publications, New York (1972)
5. V.K. Agarwal; Physics Today **41** (6), 40 (1988).
6. S. Alexander; J. Physique (Paris) **36**, 983 (1977).
7. J. Als-Nielsen, P.S. Pershan; Nucl. Instrum. Methods **208**, 545 (1983).
8. J. Als-Nielsen; Physica A **140A**, 376 (1986).
9. J. Als-Nielsen; *Structure and Dynamics of Surfaces*, Topics in Current Physics, Vol. 2, Springer, Berlin, Heidelberg (1986).
10. J. Als-Nielsen, H. Möhwald; in *Handbook of Synchrotron Radiation*, ed. by S. Ebashi, E. Rubinstein, M. Koch, Vol. 4, North-Holland (1987).
11. J. Als-Nielsen, K. Kjær; *Phase Transitions in Soft Condensed Matter*, in Proceedings of the NATO Advanced Study Institute, Geilo, Norway, ed. by T. Riste, D. Sherrington, Plenum (1989).
12. J. Als-Nielsen, D. Jacquemain, K. Kjær, F. Leveiller, M. Lahav; Phys. Rep. **246**, 251 (1994).
13. S.H. Anastasiadis, T.P. Russell, S.K. Satija, C.F. Majkrzak; Phys. Rev. Lett. **62**, 1852 (1989).
14. D. Andelmann, J.-F. Joanny, M.O. Robbins; Europhys. Lett. **7**, 731 (1988).
15. A.V. Andreev, A.G. Michette, A. Renwick; J. Modern Optics **35**, 1667 (1988).
16. D. Ausserré, A.M. Picard, L. Léger; Phys. Rev. Lett. **57**, 2671 (1986).
17. D. Bahr; PhD thesis, Kiel University (1992).
18. D. Bahr, W. Press, R. Jebasinski, S. Mantl; Phys. Rev. B **47**, 4385 (1993).
19. O. Bahr, M. Tolan, M.H. Rafailovich, J. Sokolov, J. Wang, S.K. Sinha; unpublished.
20. O. Bahr; Diplomarbeit, Kiel University (1995).
21. R.K. Balmudi, I.A. Bitsanis; J. Chem. Phys. **105**, 7774 (1996).
22. S. Banerjee, M.K. Sanyal, A. Datta, S. Kanakaraju, S. Mohan; Phys. Rev. B **54**, 16377 (1996).
23. A.L. Barabási, H.E. Stanley; *Fractal Concepts in Surface Growth*, Cambridge University Press, Cambridge, 1995.
24. T. Barberka, U. Höhne, U. Pietsch, T.H. Metzger; Thin Solid Films **244**, 1061 (1994).
25. M.F. Barnsley, R.L. Devaney, B.B. Mandelbrot, H.-O. Peitgen, D. Saupe; *The Science of Fractal Images*, Springer, Berlin, Heidelberg (1988).
26. J. Baschnagel, K. Binder; Macromolecules **28**, 6808 (1995).
27. J.K. Basu, M.K. Sanyal; Phys. Rev. Lett. **79**, 4617 (1997).

28. S. Bauer, G.R. Strobl; Polymer Bulletin **40**, 291 (1998).
29. D. Beaglehole; Phys. Rev. Lett. **58**, 1434 (1987).
30. D. Beaglehole; Physica A **200**, 696 (1990).
31. P. Beckmann, A. Spizzichino; *The Scattering of Electromagntic Waves From Rough Surfaces*, Pergamon, New York (1963).
32. J.S. Bendat, A.G. Piersol; *Random Data: Analysis and Measurement Procedures*, Wiley-Interscience, New York (1971).
33. J.S. Bendat, A.G. Piersol; *Engineering Applications of Correlation Functions and Spectral Analysis*, 2nd ed., Wiley, New York (1986).
34. B. Berne, R. Pecora; *Dynamic Light Scattering*, Wiley-Interscience, New York (1975).
35. D.W. Berreman; Phys. Rev. B **14**, 4313 (1976).
36. K. Binder; J. Chem. Phys. **79**, 6387 (1983).
37. H.-G. Birken, C. Kunz, R. Wolf; Physica Scripta **41**, 385 (1990).
38. H.-G. Birken; PhD thesis, Hamburg University (1991).
39. J. Bischof, D. Scherer, S. Herminghaus, P. Leiderer; Phys. Rev. Lett. **77**, 1536 (1996).
40. J.A.C. Bland, B. Heinrich (eds.); *Ultrathin Magnetic Structures*, Vol. 1, Springer, Berlin, Heidelberg (1994).
41. C. Blessing; PhD thesis, Hamburg University (1995).
42. K.B. Blodgett, I. Langmuir, Phys. Rev. **51**, 964 (1937).
43. S.J. Blundell, J.A.C. Bland; Phys. Rev. B **46**, 3391 (1992).
44. M. Born, E. Wolf; *Principles of Optics*, Pergamon, Oxford (1993).
45. G.E.P. Box, G.M. Jenkins; *Time Series Analysis: Forecasting and Control*, revised edition, Holden-Day, Oakland (1976).
46. J. Brandrup, E.H. Immergut; *Polymer Handbook* (1989).
47. A. Braslau, M. Deutsch, P.S. Pershan, A.H. Weiss, J. Als-Nielsen, J. Bohr; Phys. Rev. Lett. **54**, 114 (1985).
48. A. Braslau, P.S. Pershan, G. Swislow, B.M. Ocko, J. Als-Nielsen; Phys. Rev. A **38**, 2457 (1988).
49. F. Brochard-Wyart, C. Redon, C. Sykes; C. R. Acad. Sci. Ser. 2 **19**, 314 (1992).
50. D. Broseta, G.H. Fredrickson, E. Helfand, L. Leibler; Macromolecules **23**, 132 (1990).
51. L. Brügemann; PhD thesis, Kiel University (1989).
52. L. Brügemann, R. Bloch, W. Press, M. Tolan; Acta Cryst. A **48**, 688 (1992).
53. F.P. Buff, R.A. Lovett, F.H. Stillinger; Phys. Rev. Lett. **15**, 621 (1965).
54. E. Burkel; *Inelastic Scattering of X-Rays with Very High Energy Resolution*, Springer Tracts in Modern Physics, Vol. 125, Springer, Berlin, Heidelberg (1991).
55. Z.H. Cai, B. Lai, W.B. Yun, I. McNulty, K.G. Huang, T.P. Russell; Phys. Rev. Lett. **73**, 82 (1994).
56. K. Chadan, P.C. Sabatier; *Inverse Problems in Quantum Scattering Theory*, 2nd ed., Text and Monographs in Physics, Springer, Berlin, Heidelberg (1989).
57. S. Chandrasekhar; *Liquid Crystals*, 2nd ed., Cambridge University Press (1992).
58. J. Charvolin, J.-F. Joanny, J. Zinn-Justin (eds.); *Liquids at Interfaces*, Elsevier, Amsterdam (1989).
59. R. Chiarello, V. Panella, J. Krim, C. Thompson; Phys. Rev. Lett. **67**, 3408 (1991).
60. W.L. Clinton; Phys. Rev. B **48**, 1 (1993).
61. R. Cowley; Acta Cryst. A **43**, 825 (1992).
62. P. Croce, L. Névot; Revue de Physique appliquée **11**, 113 (1976).

63. C.A. Croxton; *Statistical Mechanics of the Liquid Surface*, Wiley, New York (1980).
64. H.Z. Cummins, E.R. Pike (eds.); *Photon Correlation Spectroscopy and Velocimetry*, Plenum, New York (1977).
65. J. Daillant, L. Bosio, J.J. Benattar, J. Meunier; Europhys. Lett. **8**, 453 (1989).
66. J. Daillant, J.J. Benatter, L. Léger; Phys. Rev. A **41**, 1963 (1990).
67. J. Daillant, L. Bosio, B. Harzallah, J.J. Benatter; J. Phys. II **1**, 149 (1991).
68. J. Daillant, O. Bélorgey; J. Chem. Phys. **97**, 5824 (1992).
69. J. Daillant, O. Bélorgey; J. Chem. Phys. **97**, 5837 (1992).
70. F. David; *Statistical Mechanics of Membranes and Surfaces*, ed. by D. Nelson, T. Piran, S. Weinberg, World Scientific, Singapore (1989).
71. D.K.G. de Boer; Phys. Rev. B **49**, 5817 (1994).
72. D.K.G. de Boer; Phys. Rev. B **51**, 5297 (1995).
73. D.K.G. de Boer, A.J.G. Leenaers; Physica B **221**, 18 (1996).
74. P.G. de Gennes; *Scaling Concepts in Polymer Physics*, Cornell University Press; Ithaca, New York (1979).
75. P.G. de Gennes; Macromolecules **13**, 1069 (1980).
76. P.G. de Gennes; Rev. Mod. Phys. **57**, 827 (1985).
77. P.G. de Gennes, J. Phys. Suppl. C4 **30**, 65 (1989).
78. V.O. de Haan, A.A. van Well, S. Andenwalla, G.P. Felcher; Phys. Rev. B **52**, 95 (1995).
79. V.O. de Haan, A.A. van Well, P.E. Sacks, S. Andenwalla, G.P. Felcher; Physica B **221**, 524 (1996).
80. W.H. de Jeu, J.D. Schindler, E.A.L. Mol; J. Appl. Cryst. **29**, 511 (1996).
81. W.H. de Jeu, E.A.L. Mol, G.C.L. Wong, J.-M. Petit, F. Rieutord; ESRF Newsletter **7**, 10 (1997).
82. G.B. DeMaggio, W.E. Frieze, D.W. Gidley, M. Zhu, H.A. Hristov, A.F. Yee; Phys. Rev. Lett. **78**, 1524 (1997).
83. B. Deryagin; Kolloidn. Zh. **17**, 827 (1955).
84. B.V. Deryagin, N.V. Churaev, V.M. Muller; *Surface Forces*, Consultants Bureau, New York (1987).
85. M. Deutsch, B.M. Ocko; *X-Ray and Neutron Reflectivity*, in Encyclopedia of Applied Physics Vol. 23 Wiley-VCH, 479 (1998).
86. E. Diaz-Herrera, F. Forstmann; J. Chem. Phys. **102**, 9005 (1995).
87. S.B. Dierker, R. Pindak, R.M. Fleming, I.K. Robinson, L. Berman; Phys. Rev. Lett. **75**, 449 (1995).
88. D.J. Diestler; J. Phys. Chem. **100**, 10414 (1996).
89. S. Dietrich, H. Wagner; Phys. Rev. Lett. **51**, 1469 (1983).
90. S. Dietrich, H. Wagner; Z. Phys. B **56**, 207 (1984).
91. S. Dietrich; *Wetting Phenomena*, in *Phase Transitions and Critical Phenomena*, ed. by C. Domb, J.L. Lebowitz, Vol. 12, Academic Press, New York (1988).
92. S. Dietrich, M. Napiórkowski; Physica A **177**, 437 (1991).
93. S. Dietrich; Physica Scripta **49**, 519 (1993).
94. S. Dietrich, A. Haase; Phys. Rep. **260**, 1 (1995).
95. S. Dietrich; J. Phys. Cond. Matter **8**, 9127 (1996).
96. A.K. Doerr, M. Tolan, T. Seydel, W. Press; Physica B **248**, 263 (1998).
97. A.K. Doerr, X.Z. Wu, B.M. Ocko, E.B. Sirota, O. Gang, M. Deutsch; Colloids Surfaces A: Physicochem. Eng. Aspects **128**, 63 (1997).
98. A.K. Doerr, M. Tolan, J.-P. Schlomka, T. Seydel, W. Press; unpublished.
99. A.K. Doerr, PhD thesis, Kiel University (1999).
100. M. Doi; *Introduction to Polymer Physics*, Oxford Science Publications, Clarendon Press (1996).

101. H. Dosch, B.W. Batterman, D.C. Wack; Phys. Rev. Lett. **56**, 1144 (1986).
102. H. Dosch; Phys. Rev. B **35**, 2137 (1987).
103. H. Dosch; *Critical Phenomena at Surfaces and Interfaces (Evanescent X-Ray and Neutron Scattering)*, Springer Tracts in Modern Physics, Vol. 126, Springer, Berlin, Heidelberg (1992).
104. S.M. Durbin; Acta Cryst. A **51**, 258 (1995).
105. P. Dutta, S.K. Sinha; Phys. Rev. Lett. **47**, 50 (1981).
106. I.E. Dzyaloshinskii, E.M. Lifshitz, L.P. Pitaevskii; Adv. Phys. **10**, 165 (1961).
107. C. Ebner, W.F. Saam, Phys. Rev. Lett. **38**, 1486 (1977).
108. S.F. Edwards, D.R. Wilkinson; Proc. R. Soc. Lond. **381**, 17 (1982).
109. J. Eggebrecht, K.E. Gubbins, S.M. Thompson; J. Chem. Phys. **86**, 2286 (1987).
110. J. Eggebrecht, S.M. Thompson, K.E. Gubbins; J. Chem. Phys. **86**, 2299 (1987).
111. U. Englisch, T. Gutberlet, R. Steitz, R. Oeser, U. Pietsch; phys. stat. sol. (b) **201**, 67 (1997).
112. T. Engøy, K.J. Måløy, A. Hansen, S. Roux; Phys. Rev. Lett. **73**, 834 (1994).
113. R. Evans; Adv. Phys. **28**, 143 (1979).
114. F. Faupel, R. Willecke, A. Thran, M. Kiene, C. von Bechtolsheim, T. Strunskus; Defect and Diffusion Forum **143–147**, 887 (1997).
115. R. Feidenhans'l; Surface Science Reports **10**, 105 (1989).
116. G.P. Felcher; Phys. Rev. B **24**, 1595 (1981).
117. G.P. Felcher; Physica B **198**, 150 (1994).
118. G.P. Felcher, T.P. Russell (eds.); Proceedings of the workshop on "Methods of Analysis and Interpretation of Neutron Reflectivity Data", Physica B **173** (1991).
119. Y.P. Feng, S.K. Sinha, H.W. Deckmann, J.B. Hastings, D.P. Siddons; Phys. Rev. Lett. **71**, 537 (1993).
120. Y.P. Feng, H.W. Deckmann, S.K. Sinha; Appl. Phys. Lett. **64**, 930 (1994).
121. Y.P. Feng, C.F. Majkrzak, S.K. Sinha, D.G. Wiesler, H. Zhang, H.W. Deckmann; Phys. Rev. B **49**, 10814 (1994).
122. E. Findeisen; Diplomarbeit, Kiel University (1992).
123. E. Findeisen, L. Brügemann, J. Stettner, M. Tolan; J. Phys.: Condens. Matter **5**, 8149 (1993).
124. S. Fisk, B. Widom; J. Chem. Phys. **50**, 3219 (1969).
125. G. Flöter, S. Dietrich; Z. Phys. B **97**, 213 (1995).
126. J.A. Forrest, K. Dalnoki-Veress, J.R. Stevens, J.R. Dutcher; Phys. Rev. Lett. **77**, 2002 (1996).
127. C. Fradin, A. Braslau, D. Luzet, M. Alba, C. Gourier, J. Daillant, G. Grübel, G. Vignaud, J.F. Legrand, J. Lal, J.M. Petit, F. Rieutord; Physica B **248**, 310 (1998).
128. G.H. Fredrickson, A. Ajdari, L. Leibler, J.-P. Carton; Macromolecules **25**, 2882 (1992).
129. Y.L. Frenkel; *Kinetic Theory of Liquids*, Clarendon Press, Oxford (1946).
130. P. Frodl, S. Dietrich; Physica A **45**, 7330 (1992).
131. P. Frodl, S. Dietrich; Phys. Rev. E **48**, 3741 (1993).
132. H. Fuchs, H. Ohst, W. Prass; Adv. Mater. **3**, 10 (1991).
133. E.E. Fullerton, J. Pearson, C.H. Sowers, S.D. Bader, X.Z. Wu, S.K. Sinha; Phys. Rev. B **48**, 17432 (1993).
134. M. Gailhanou, G.T. Baumbach, U. Marti, P.C. Silva, F.K. Reinhart, M. Ilegems; Appl. Phys. Lett. **62**, 1623 (1993).
135. S. Garoff, E.B. Sirota, S.K. Sinha, H.B. Stanley; J. Chem. Phys. **90**, 7505 (1989).

136. E.L. Gartstein, R.A. Cowley; Acta Cryst. A **46**, 576 (1990).
137. C. Gerthsen, H.O. Kneser, H. Vogel; *Physik*, 16th ed., Springer, Berlin, Heidelberg (1990).
138. M. Ghosh, K.L. Mittal (eds.); *Polyimids: Fundamental Aspects and Technological Applications*, Marcel Dekker, New York (1996).
139. A. Gibaud, G. Vignaud, S.K. Sinha; Acta Cryst. A **49**, 642 (1993).
140. A. Gibaud, N. Cowlam, G. Vignaud, T. Richardson; Phys. Rev. Lett. **74**, 3205 (1995).
141. A. Gibaud, J. Wang, M. Tolan, G. Vignaud, S.K. Sinha; J. de Physique I **6**, 1085 (1996).
142. S. Gierlotka, P. Lambooy, W.H. de Jeu; Europhys. Lett. **12**, 341 (1990).
143. T. Gog, P.M. Len, D. Bahr, G. Materlik, C.S. Fadley, C. Sanchez-Hanke; Phys. Rev. Lett. **76**, 3132 (1996).
144. B. Götzelmann, A. Haase, S. Dietrich, Phys. Rev. E **53**, 3456 (1996).
145. M. Goulian, N. Lei, J. Miller, S.K. Sinha; Phys. Rev. A **46**, R6170 (1992).
146. C. Gourier, J. Daillant, A. Braslau, M. Alba, K. Quinn, D. Luzet, C. Blot, D. Chatenay, G. Grübel, J.-F. Legrand, G. Vignaud; Phys. Rev. Lett. **78**, 3157 (1997).
147. I.S. Gradshteyn, I.M. Ryzhik; *Tables of Integrals, Series and Products*, ed. by A. Jeffrey, Academic Press, London (1980).
148. G. Grübel, J. Als-Nielsen, D. Abernathy, G. Vignaud, S. Brauer, G.B. Stephenson, S.G.J. Mochrie, M. Sutton, I.K. Robinson, R. Fleming, R. Pindak, S. Dierker, J.F. Legrand; ESRF Newsletter **20** (1994).
149. G. Grübel; *Spectroscopy with Coherent X-Rays*, A-FEL workshop, Hamburg (1996).
150. B. Guckenbiehl, M. Stamm, T. Springer; Physica B **198**, 127 (1994).
151. S. Gupta, D.C. Koopman, G.B. Westermann-Clark, I.A. Bitsanis; J. Chem. Phys. **100**, 8444 (1994).
152. H.C. Hamaker; Physica **4**, 1058 (1937).
153. W.A. Hamilton, R. Pynn; Physica B **173**, 71 (1991).
154. W. Hansen, J.P. Kotthaus, U. Merkt; Semiconductors and Semimetals – Nanostructured Systems, Vol. 35, Academic Press, London, San Diego, 279 (1992).
155. J.L. Harden, H. Pleiner, P.A. Pincus; J. Chem. Phys. **94**, 5208 (1991).
156. D. Heitmann, J.P. Kotthaus; Physics Today **46**(6) 56 (1993).
157. E. Helfand, Y. Tagami; J. Chem. Phys. **56**, 3592 (1972).
158. E. Helfand, S.M. Bhattacharjee, G.H. Fredrickson; J. Chem. Phys. **91**, 7200 (1972).
159. W. Helfrich; Z. Naturforsch. c **28**, 693 (1973).
160. W. Helfrich; Z. Naturforsch. a **33**, 305(1978).
161. K. Hermansson, U. Lindberg, B. Hök, G. Palmskog; Proc. IEEE **91**, 193 (1991).
162. F. Holovko, E.V. Vakarin; Molecular Physics **87**, 1375 (1996).
163. V. Holý, L. Tapfer, E. Koppensteiner, G. Bauer, H. Lage, O. Brandt, K. Ploog; Appl. Phys. Lett. **63**, 3140 (1993).
164. V. Holý, J. Kuběna, I. Ohlídal, K. Lischka, W. Plotz; Phys. Rev. B **47**, 15896 (1993).
165. V. Holý, T. Baumbach; Phys. Rev. B **49**, 10668 (1994).
166. V. Holý, C. Giannini, L. Tapfer, T. Marschner, W. Stolz; Phys. Rev. B **55**, 9960 (1997).
167. V. Holý, U. Pietsch, T. Baumbach; *X-Ray Scattering from Thin Films (High Resolution X-Ray Scattering from Crystalline Thin Films)*, Springer Tracts in Modern Physics, Vol. 149, Springer, Berlin, Heidelberg (1998).

168. R. Holyst; Phys. Rev. A **44**, 3692 (1991).
169. S. Iatsevitch, F. Forstmann; J. Chem. Phys. **107**, (1997).
170. J. Israelachvili; *Intramolecular Surface Forces* Second Edition, Academic Press, London, San Diego (1992).
171. J. Israelachvili; Langmuir **10**, 3774 (1994).
172. J. Jäckle, K. Kawasaki; J. Phys.: Condens. Matter **7**, 4351 (1995).
173. J. Jäckle; private communication (1997).
174. J. Jäckle; J. Phys.: Condens. Matter **10**, 7121 (1998).
175. K. Jacobs, S. Herminghaus, K.R. Mecke; Langmuir **14**, 965 (1998).
176. K. Jacobs, R. Seemann, G. Schatz, S. Herminghaus; Langmuir Lett. (1998) in press.
177. K. Jacobs, R. Seemann, G. Schatz, S. Herminghaus; preprint.
178. R.W. James; *The Optical Principles of the Diffraction of X-Rays*, Ox Bow Press, Woodbridge, Connecticut (1982).
179. W. Jark, S. Di Fonzo, S. Lagomarsino, A. Cedola, E. Di Fabrizio, A. Bram, C. Riekel; J. Appl. Phys. **80**, 4831 (1996).
180. Y.C. Jean, R. Zhang, H. Cao, J.-P.Yuan, C.-M. Huang, B. Nielsen, P. Asoka-Kumar; Phys. Rev. B **56**, R8459 (1997).
181. K. Jeß, V. Helbig, J. Kaspareit, V. Rhode, T. Weirauch; J. Quant. Spectr. Radiat. Transf. **1** (1990).
182. M. Kardar, G. Parisi, Y.-C. Zhang; Phys. Rev. Lett. **56**, 889 (1986).
183. M. Kasch, F. Forstmann; J. Chem. Phys. **99**, 3037 (1993).
184. J.L. Keddie, R.A.L Jones, R.A. Cory; Europhys. Lett. **27**, 59 (1994).
185. J.L. Keddie, R.A.L Jones, R.A. Cory; Faraday Discuss. **98**, 219 (1994).
186. W. Kendall; in *Kendall's Advanced Theory of Statistics*, Vol. 1, 6th ed., ed. by A. Stuart, J.K. Ord, Edward Arnold, London (1994).
187. D.A. Kessler, H. Levine, L.M. Sander; Phys. Rev. Lett. **69**, 100 (1992).
188. H. Kiessig; Annalen der Physik **10**, 769 (1931).
189. U. Klemradt; PhD thesis, RWTH Aachen (1994).
190. U. Klemradt, M. Funke, M. Fromm, B. Lengeler, J. Peisl, A. Förster; Physica B **221**, 27 (1996).
191. U. Klemradt, M. Fromm, G. Landmesser, H. Amschler, J. Peisl; Physica B **248**, 83 (1998).
192. M.V. Klibanov, P.E. Sacks; J. Comput. Phys. **112**, 273 (1994).
193. E.A. Kondrashkina, S.A. Stepanov, R. Opitz, M. Schmidbauer, R. Köhler, R. Hey, M. Wassermeier, D.V. Novikov; Phys. Rev. B **56**, 10469 (1997).
194. J.B. Kortright; J. Appl. Phys. **70**, 3620 (1991).
195. P. Kosmol; *Methoden zur numerischen Behandlung nichtlinearer Gleichungen und Optimierungsaufgaben*, BG Teubner Studienbücher, Stuttgart (1989).
196. J. Krim, J. Suzanne, H. Shechter, R. Wang, H. Taub; Surf. Sci. **165**, 446 (1985).
197. J. Krim, G. Palasantzas; Int. J. Mod. Phys. B **9**, 599 (1995).
198. J. Krug; Adv. Phys. **46**, 139 (1997).
199. K. Kunz, M. Stamm; Macromol. Symp. **78**, 105 (1994).
200. K. Kunz, M. Stamm; Macromolecules **29**, 2548 (1996).
201. S. Lagomarsino, A. Cedola, P. Cloetens, S. Di Fonzo, W. Jark, G. Soullié, C. Riekel; Appl. Phys. Lett. **71**, 3 (1997).
202. P. Lambooy, S. Gierlotka, I.W. Hamley, W.H. de Jeu; in *Phase Transitions in Liquid Crystals*, ed. by S. Martellucci, A.N. Chester, Plenum, New York, 239 (1992).
203. D. Langevin (ed.); *Light Scattering by Liquid Surfaces and Complementary Techniques*, Marcel Dekker, New York (1992).
204. J.C. Lee; Phys. Rev. B **46**, 8648 (1992).

205. L. Léger, J.F. Joanny; Rep. Prog. Phys. **55**; 431 (1992).
206. J. Lekner; *Theory of Reflection*, Martinus Nijhoff, Dordrecht Boston, Lancaster (1987).
207. J. Lekner; Physica B **173**, 99 (1991).
208. H. Li, M. Kardar; Phys. Rev. B **42**, 6546 (1990).
209. Z. Li, W. Zhao, J. Quinn, M.H. Rafailovich, J. Sokolov, R.B. Lennox, A. Eisenberg, X.Z. Wu, M.W. Kim, S.K. Sinha, M. Tolan; Langmuir **11**, 4785 (1995).
210. Z. Li; PhD thesis, SUNY Stony Brook (1996).
211. Z. Li, S. Qu, M.H. Rafailovich, J. Sokolov, M. Tolan, M.S. Turner, J. Wang, S.A. Schwarz, H. Lorenz, J.P. Kotthaus; Macromolecules **30**, 8410 (1997).
212. Z. Li, M. Tolan, T. Höhr, D. Kharas, S. Qu, J. Sokolov, M.H. Rafailovich, H. Lorenz, J.P. Kotthaus, J. Wang, S.K. Sinha, A. Gibaud; Macromolecules **31**, 1915 (1998).
213. Z.X. Li, P.N. Thirtle, R.K. Thomas, J. Penfold, J.R.P. Webster, A.R: Rennie; Physica B **248**, 171 (1998).
214. J.L. Libbert, J.A. Pitney, I.K. Robinson; J. Synchrotron Rad. **4**, 125 (1997).
215. J.L. Libbert, R. Pindak, S.B. Dierker, I.K. Robinson; Phys. Rev. B **56**, 6454 (1997).
216. A. Lied, H. Dosch, J. Bilgram; Phys. Rev. Lett. **72**, 3884 (1994).
217. B. Lin, M.L Schlossman, M. Meron, S.M. Williams, P.J. Viccaro; Rev. Sci. Instrum. **66**, 1 (1995).
218. R. Lipowsky; Phys. Rev. B **32**, 1731 (1985).
219. R. Lipperheide, G. Reiss, H. Fiedeldey, S.A. Sofianos, H. Leeb; Phys. Rev. B **51**, 11032 (1995).
220. R. Lipperheide, G. Reiss, H. Leeb, S.A. Sofianos; Physica B **221**, 514 (1996).
221. R. Lipperheide, H. Leeb; Physica B **248**, 366 (1998).
222. A.J. Liu, M.E. Fisher; Phys. Rev. A **40**, 7202 (1989).
223. R. Lovett, C.Y. Mou, F.P. Buff; J. Chem. Phys. **65**, 570 (1976).
224. C.A. Lucas, E. Gartstein, R.A. Cowley; Acta Cryst. A **45**, 416 (1989).
225. L.B. Lurio, T.A. Rabedeau, P.S. Pershan, I.F. Silvera, M. Deutsch, S.D. Kosowsky, B.M. Ocko; Phys. Rev. Lett. **68**, 2628 (1992).
226. L.B. Lurio, T.A. Rabedeau, P.S. Pershan, I.F. Silvera, M. Deutsch, S.D. Kosowsky, B.M. Ocko; Phys. Rev. B **48**, 9644 (1993).
227. M. Lütt, J.-P. Schlomka, M. Tolan, J. Stettner, O.H. Seeck, W. Press; Phys. Rev. B **56**, 4085 (1997).
228. M. Lütt, M.R. Fitzsimmons, D.Q. Li; J. Phys. Chem. B **102**, 400 (1998).
229. J.J. Magda, M. Tirrell, H.T. Davis; J. Chem. Phys. **83**, 1888 (1985).
230. O.M. Magnussen, B.M. Ocko, M.J. Regan, K. Penanen, P.S Pershan, M. Deutsch; Phys. Rev. Lett. **74**, 4444 (1995).
231. S. Majaniemi, T. Ala-Nissila, J. Krug; Phys. Rev. B **53**, 8071 (1996).
232. C.F. Majkrzak, N.F. Berk; Physica B **221**, 520 (1996).
233. C.F. Majkrzak, N.F. Berk, J.A. Dura, S.K. Satija, A. Karim, J. Pedulla, R.D. Deslattes; Physica B **248**, 338 (1998).
234. L. Mandel, E. Wolf; *Optical Coherence and Quantum Optics*, Cambridge University Press (1995).
235. B.B. Mandelbrot; *The Fractal Geometry of Nature*, W.H. Freeman, New York (1982).
236. K.F. Mansfield, D.N. Theodorou; Macromolecules **24**, 6283 (1991).
237. P. Martin, A. Buguin, F. Brochard-Wyart; Europhys. Lett. **28**, 421 (1994).
238. A.M. Mayes; Macromolecules **29**, 3114 (1994).
239. B.R. McClain, D.D. Lee, B.L. Carvalho, S.G.J. Mochrie, S.H. Chen, J.D. Lister; Phys. Rev. Lett. **72**, 246 (1994).
240. B.R. McClain, M. Yoon, J.D. Lister, S.G.J. Mochrie; preprint.

241. K.R. Mecke, S. Dietrich; preprint.
242. A. Messiah, *Quantenmechanik*, Vol. 2, 2nd ed., Walter de Gruyter, Berlin, New York (1985).
243. J. Meunier; J. Physique (Paris) **48**, 1819 (1987).
244. Z.H. Ming, A. Krol, Y.L. Soo, Y.H. Kao, J.S. Park, K.L. Wang; Phys. Rev. B **47**, 16373 (1993).
245. S.G.J. Mochrie, A.M. Mayes, A.R. Sandy, M. Sutton, S. Brauer, G.B. Stephenson, D.L. Abernathy, G. Grübel; Phys. Rev. Lett. **78**, 1275 (1997).
246. E.A.L. Mol, J.D. Shindler, A.N. Shalaginov, W.H. de Jeu; Phys. Rev. E **54**, 536 (1996).
247. E.A.L. Mol; PhD thesis, FOM Institute (AMOLF), Amsterdam (1997).
248. E.A.L. Mol, G.C.L. Wong, J.-M. Petit, F. Rieutord, W.H. de Jeu; Phys. Rev. Lett. **79**, 3439 (1997).
249. E.A.L. Mol, G.C.L. Wong, J.M. Petit, F. Rieutord, W.H. de Jeu; Physica B **248**, 191 (1998).
250. P. Müller-Buschbaum, M. Tolan, W. Press; Z. Phys. B **95**, 331 (1994).
251. P. Müller-Buschbaum, M. Tolan, W. Press, F. Brinkop, J.P. Kotthaus; Berichte der Bunsengesellschaft **98**, 413 (1994).
252. P. Müller-Buschbaum, M. Strzelczyk, M. Tolan, W. Press; Z. Phys. B **98**, 89 (1995).
253. P. Müller-Buschbaum, P. Vanhoorne, V. Scheumann, M. Stamm; Europhys. Lett. **40**, 655 (1997).
254. P. Müller-Buschbaum, S.A. O'Neill, S. Affrossman, M. Stamm; Macromolecules **31**, 3686 (1998).
255. P. Müller-Buschbaum, M. Casagrande, J. Gutmann, T. Kuhlmann, M. Stamm, G. von Krosigk, U. Lode, S. Cunis, R. Gehrke; Europhys. Lett. **42**, 517 (1998).
256. M. Napiórkowski, S. Dietrich; Z. Phys. B **89**, 263 (1992).
257. M. Napiórkowski, S. Dietrich; Phys. Rev. E **47**, 1836 (1993).
258. M. Napiórkowski; Berichte der Bunsengesellschaft **98**, 352 (1994).
259. M. Napiórkowski, S. Dietrich; Z. Phys. B **97**, 511 (1995).
260. H.B. Neumann; Diplomarbeit, Hamburg University (1991).
261. L. Névot, P. Croce; Revue de Physique appliquée **15**, 761 (1980).
262. J.C. Newton; PhD thesis, University of Missouri-Columbia (1989).
263. V. Nitz; PhD thesis, Kiel University (1995).
264. V. Nitz, M. Tolan, J.-P. Schlomka, O.H. Seeck, J. Stettner, W. Press, M. Stelzle, E. Sackmann; Phys. Rev. B **54**, 5038 (1996).
265. K.A. Nugent; J. Opt. Soc. Am. A **8**, 1574 (1991).
266. K.A. Nugent, J.E. Trebes; Rev. Sci. Instrum. **63**, 2146 (1992).
267. B.M. Ocko, X.Z. Wu, E.B. Sirota, S.K. Sinha, M. Deutsch; Phys. Rev. Lett. **72**, 242 (1994).
268. T. Ohkawa, Y. Yamaguchi, O. Sakata, M.K. Sanyal, A. Datta, S. Banerjee, H. Hashizume; Physica B **221**, 416 (1996).
269. R. Opitz; PhD thesis, Humboldt University, Berlin (1998).
270. W.J. Orts, J.H. van Zanten, W.-L. Wu, S.K. Satija; Phys. Rev. Lett. **71**, 867 (1993).
271. G. Palasantzas, J. Krim; Phys. Rev. B **48**, 2873 (1993).
272. G. Palasantzas; Phys. Rev. B **48**, 14472 (1993).
273. G. Palasantzas; Phys. Rev. B **49**, 10544 (1994).
274. R. Paniago, H. Homma, P.C. Chow, S.C. Moss, Z. Barnea, S.S.P. Parkin, D. Cookson; Phys. Rev. B **52**, 17502 (1995).
275. R. Paniago, H. Homma, P.C. Chow, H. Reichert, S.C. Moss, Z. Barnea, S.S.P. Parkin, D. Cookson; Physica B **221**, 10 (1996).
276. R. Paproth; Diplomarbeit, Kiel University (1998).

277. L.G. Parratt; Phys. Rev. **95**, 359 (1954).
278. C.N. Patra, S.K. Ghosh; J. Chem. Phys. **106**, 2763 (1997).
279. D. Perahia, D.G. Wiesler, S.K. Satija, L.J. Fetters, S.K. Sinha, S.T. Milner; Phys. Rev. Lett. **72**, 100 (1994).
280. P.S. Pershan; Faraday Discuss. Chem. Soc. **89**, 231 (1990).
281. P.S. Pershan; Physica A **231**, 111 (1997).
282. Y.-H. Phang, R. Kariotis, D.E. Savage, M.G. Lagally; J. Appl. Phys. **72**, 4627 (1992).
283. J. Picht; Ann. Phys. **5**, 433 (1929).
284. Z.G. Pinsker; *Dynamical Scattering of X-Rays in Crystals*, Springer Series in Solid State Science Vol. 3, Springer, Berlin, Heidelberg (1978).
285. A. Plech, U. Klemradt, H. Metzger, J. Peisl; J. Phys.: Condens. Matter **10**, 971 (1998).
286. M. Plischke, D. Henderson; J. Chem. Phys. **84**, 2846 (1986).
287. W. Press, M. Tolan, J. Stettner, V. Nitz, J.-P. Schlomka, O.H. Seeck, P. Müller-Buschbaum; Physica B **221**, 1 (1996).
288. W. Press, J.-P. Schlomka, M. Tolan, B. Asmussen; J. Appl. Cryst. **30**, 963 (1997).
289. P.R. Pukite, C.S. Lent, P.I. Cohen; Surf. Sci. **161**, 39 (1985).
290. R. Pynn; Phys. Rev. B **45**, 602 (1992).
291. S. Qu; PhD thesis, SUNY Stony Brook (1997).
292. C. Redon, F. Brochard-Wyart, F. Rondelez; Phys. Rev. Lett. **66**, 715 (1991).
293. M.J. Regan, E.H. Kawamoto, S. Lee, P.S Pershan, N. Maskil, M. Deutsch, O.M. Magnussen, B.M. Ocko, L.E. Bermann; Phys. Rev. Lett. **75**, 2498 (1995).
294. M.J. Regan, H. Tostmann, P.S Pershan, O.M. Magnussen, E. DiMasi, B.M. Ocko, M. Deutsch; Phys. Rev. B **55**, 10786 (1997).
295. G. Reiter; Phys. Rev. Lett. **68**, 75 (1992).
296. G. Reiss; Physica B **221**, 533 (1996).
297. F. Rieutord, A. Braslau, R. Simon, H.J. Lauter, V. Pasyuk; Physica B **221**, 538 (1996).
298. M.O. Robbins, D. Andelmann, J.-F. Joanny; Phys. Rev. A **43**, 4344 (1991).
299. I.K. Robinson, R. Pindak, R.M. Fleming, S.B. Dierker, K. Ploog, G. Grübel, D.L. Abernathy, J. Als-Nielsen; Phys. Rev. B **52**, 9917 (1995).
300. I.K. Robinson, J.A. Pitney, J.L. Libbert, I.A. Vartanyants; Physica B **248**, 387 (1998).
301. P. Roche, G. Deville, K.O. Keshishev, N.J. Appleyard, F.I.B. Williams; Phys. Rev. Lett. **75**, 3316 (1995).
302. G.T. Ruck, D.E. Barrick, W.D. Stuart, C.K. Krichbaum; *Radar Cross Section Handbook*, Plenum, New York (1970).
303. T.P. Russell; Mater. Sci. Rep. **5**, 171 (1990).
304. T.P. Russell; Physica B **221**, 267 (1996).
305. S.A. Safran, J. Klein; J. Phys. II, France **3**, 749 (1993).
306. T. Salditt, T.H. Metzger, J. Peisl; Phys. Rev. Lett. **73**, 2228 (1994).
307. T. Salditt, H. Rhan, T.H. Metzger, J. Peisl, R. Schuster, J.P. Kotthaus; Z. Phys. B **96**, 227 (1994).
308. T. Salditt, T.H. Metzger, J. Peisl, X. Jiang; J. Phys. III France **4**, 1573 (1994).
309. T. Salditt; PhD thesis, LMU München (1995).
310. T. Salditt, T.H. Metzger, J. Peisl, R. Reinker, M. Moske, K. Samwer; Europhys. Lett. **32**, 331 (1995).
311. T. Salditt, T.H. Metzger, J. Peisl, G. Goerigk; J. Phys. D: Appl. Phys. **28**, A236 (1995).
312. T. Salditt, D. Lott, T.H. Metzger, J. Peisl, G. Vignaud, J.F. Legrand, G. Grübel, P. Høghøi, O. Schärpf; Physica B **221**, 13 (1996).

313. M.K. Sanyal, S.K. Sinha, K.G. Huang, B.M. Ocko; Phys. Rev. Lett. **66**, 628 (1991).
314. M.K. Sanyal, S.K. Sinha, A. Gibaud, S.K. Satija, C.F. Majkrzak, H. Homma; in *Surface X-Ray and Neutron Scattering*, Springer Proceedings in Physics, Vol. 61, ed. by H. Zabel, I.K. Robinson, Springer, Berlin, Heidelberg (1992) p. 91.
315. M.K. Sanyal, S.K. Sinha, A. Gibaud, K.G. Huang, B.L. Carvalho, M. Rafailovich, J. Sokolov, X. Zhao, W. Zhao; Europhys. Lett. **21**, 691, (1993).
316. M.K. Sanyal, J.K. Basu, A. Datta, S. Banerjee; Europhys. Lett. **36**, 265 (1996).
317. M.K. Sanyal, J.K. Basu, A. Datta; Physica B **248**, 217 (1998).
318. M.K. Sanyal, A. Datta, A.K. Srivastava, B.M. Arora, S. Banerjee, P. Chakraborty, F. Caccavale, O. Sakata, H. Hashizume; unpublished.
319. M.K. Sanyal, S. Hazra, J.K. Basu, A. Datta; (1998) Preprint.
320. D.E. Savage, N. Schimke, Y.-H. Phang, M.G. Lagally; J. Appl. Phys. **71**, 3283 (1992).
321. D.E. Savage, Y.-H. Phang, J.J. Rownd, J.F. MacKay, M.G. Lagally; J. Appl. Phys. **74**, 6158 (1993).
322. A.C. Schell; PhD thesis, MIT, Cambridge, Mass. (1961).
323. J.-P. Schlomka, M. Tolan, L. Schwalowsky, O.H. Seeck, J. Stettner, W. Press; Phys. Rev. B **51**, 2311 (1995).
324. J.-P. Schlomka, M.R. Fitzsimmons, R. Pynn, J. Stettner, O.H. Seeck, M. Tolan, W. Press; Physica B **221**, 44 (1996).
325. D.W. Schubert, M. Stamm; Europhys. Lett. **35**, 419 (1996).
326. D.W. Schubert; PhD thesis, Mainz University (1996).
327. D.W. Schubert; Polymer Bulletin **38**, 177 (1997).
328. R. Schwedhelm, J.-P. Schlomka, R. Adelung, S. Woedtke, L. Kipp, M. Tolan, W. Press, M. Skibowski; preprint.
329. O.H. Seeck; Diploma thesis, Kiel University (1994).
330. O.H. Seeck, P. Müller-Buschbaum, M. Tolan, W. Press; Europhys. Lett. **29**, 699 (1995).
331. T. Seydel, M. Tolan, W. Prange, A. Doerr, J.-P. Schlomka, W. Press, M.H. Rafailovich, J. Sokolov; unpublished.
332. S. Sferrazza, C. Xiao, R.A.L Jones, D.G. Bucknall, J. Webster, J. Penfold; Phys. Rev. Lett. **78**, 3693 (1997).
333. Q. Shen, C.C. Umbach, B. Weselak, J.M. Blakely; Phys. Rev. B **48**, 17967 (1993).
334. J.D. Shindler, E.A.L. Mol, A. Shalaginov, W.H. de Jeu; Phys. Rev. Lett. **74**, 722 (1995).
335. S.K. Sinha, E.B. Sirota, S. Garoff, H.B. Stanley; Phys. Rev. B **38**, 2297 (1988).
336. S.K. Sinha, M.K. Sanyal, K.G. Huang, A. Gibaud, M. Rafailovich, J. Sokolov, X. Zhao, W. Zhao; in *Surface X-Ray and Neutron Scattering*, Springer Proceedings in Physics, Vol. 61, ed. by H. Zabel, I.K. Robinson, Springer, Berlin, Heidelberg (1992) p. 85.
337. S.K. Sinha, M.K. Sanyal, S.K. Satija, C.F. Majkrzak, D.A. Neumann, H. Homma, S. Szpala, A. Gibaud, H. Morkoc; Physica B **198**, 72 (1994).
338. S.K. Sinha; J. Phys. III, France **4**, 1543 (1994).
339. S.K. Sinha, M. Tolan, G. Vacca, Z. Li, M. Rafailovich, J. Sokolov, H. Lorenz, J.P. Kotthaus, Y.P. Feng, G. Grübel, D. Abernathy; in *"Dynamics in Small Confining Systems II"*, Vol. 366, ed. by J.M. Drake, J. Klafter, R. Kopelman, S.M. Toian, Materials Research Society, Boston (1995) p. 3.
340. S.K. Sinha, M. Tolan, A. Gibaud; Phys. Rev. B **57**, 2740 (1998).
341. E.B. Sirota, H.E. King, D.M. Singer, H.H. Shao; J. Chem. Phys. **98**, 5809 (1993).

342. D.S. Sivia, W.A. Hamilton, G.S. Smith, T.P. Riecker, R. Pynn; J. Appl. Phys. **70**, 732 (1991).
343. D.S. Sivia; *Data Analysis: A Bayesian Tutorial*, Oxford University Press (1996).
344. D.S. Sivia, J.R.P. Webster; Physica B **248**, 327 (1998).
345. E. Spiller, A. Segmüller; Appl. Phys. Lett. **24**, 60 (1974).
346. E. Spiller, D. Stearns, M. Krumrey; J. Appl. Phys. **74**, 107 (1993).
347. M. Stamm, S. Hüttenbach, G. Reiter, T. Springer; Europhys. Lett. **14**, 451 (1991).
348. M. Stamm; *Reflection of Neutrons for the Investigation of Polymer Interdiffusion at Interfaces*, Physics of Polymer Surfaces and Interfaces, ed. by I.C. Sanchez, Butterworth-Heinemann, Boston, 163 (1992).
349. M. Stamm, D.W. Schubert; Ann. Rev. Mater. Sci. **25**, 325 (1995).
350. F. Stanglmeier; PhD thesis, RWTH Aachen (1990).
351. F. Stanglmeier, B. Lengeler, W. Weber, G. Göbel, M. Schuster; Acta Cryst. A **48**, 626 (1992).
352. D.G. Stearns; J. Appl. Phys. **71**, 4286 (1992).
353. M. Stelzle; PhD thesis, TU München (1992).
354. J. Stettner; PhD thesis, Kiel University (1995).
355. J. Stettner, L. Schwalowsky, O.H. Seeck, M. Tolan, W. Press, C. Schwarz, H. von Känel; Phys. Rev. B **53**, 1398 (1996).
356. R. Stömmer, J. Grenzer, J. Fischer, U. Pietsch; J. Phys. D: Appl. Phys. **28**, A216 (1995).
357. R. Stömmer, U. English, U. Pietsch, V. Holý; Physica B **221**, 284 (1996).
358. R. Stömmer, U. Pietsch; J. Phys. D: Appl. Phys. **29**, 3161 (1996).
359. S. Stringari, J. Treiner; Phys. Rev. B **36**, 8369 (1987).
360. M. Strzelczyk, P. Müller-Buschbaum, M. Tolan, W. Press; Phys. Rev. B **52**, 16869 (1995).
361. M. Sutton, S.G.J. Mochrie, T. Greytak, S.E. Nagler, L.E. Berman, G.A. Held, G.B. Stephenson; Nature (London) **352**, 608 (1991).
362. M. Takeda, Y. Endoh, A. Kamijo, J. Mizuki; Physica B **248**, 14 (1998).
363. H. Tang, K.F. Freed; J. Chem. Phys. **94**, 6307 (1991).
364. T. Thurn-Albrecht, W. Steffen, A. Patkowski, G. Meier, E.W. Fischer, G. Grübel, D.L. Abernathy; Phys. Rev. Lett. **77**, 5437 (1996).
365. I.M. Tidswell, T.A. Rabedeau, P.S. Pershan, S.D. Kosowsky; Phys. Rev. Lett. **66**, 2108 (1991).
366. I.M. Tidswell, T.A. Rabedeau, P.S. Pershan, J.P. Folkers, M.V. Baker, G.M. Whitesides; Phys. Rev. B **44**, 10869 (1991).
367. M. Tolan, G. König, L. Brügemann, W. Press, F. Brinkop, J.P. Kotthaus; Europhys. Lett. **20**, 223 (1992).
368. M. Tolan, W. Press, F. Brinkop, J.P. Kotthaus; J. Appl. Phys. **75**, 7761 (1994).
369. M. Tolan, D. Bahr, J. Süßenbach, W. Press, F. Brinkop, J.P. Kotthaus; Physica B **198**, 55 (1994).
370. M. Tolan, W. Press, F. Brinkop, J.P. Kotthaus; Phys. Rev. B **51**, 2239 (1995).
371. M. Tolan, G. Vacca, S.K. Sinha, Z. Li, M. Rafailovich, J. Sokolov, H. Lorenz, J.P. Kotthaus; J. Phys. D: Appl. Phys. **28**, A231 (1995).
372. M. Tolan, G. Vacca, J. Wang, S.K. Sinha, Z. Li, M. Rafailovich, J. Sokolov, A. Gibaud, H. Lorenz, J.P. Kotthaus; Physica B **221**, 53 (1996).
373. M. Tolan, G. Vacca, S.K. Sinha, Z. Li, M.H. Rafailovich, J. Sokolov, H. Lorenz, J.P. Kotthaus; Appl. Phys. Lett. **68**, 191 (1996).
374. M. Tolan, W. Press; Z. Kristallogr. **213**, 319 (1998).
375. M. Tolan, S.K. Sinha; Physica B **248**, 399 (1998).

376. M. Tolan, O.H. Seeck, J.-P. Schlomka, W. Press, J. Wang, S.K. Sinha, Z. Li, M.H. Rafailovich, J. Sokolov; Phys. Rev. Lett. **81** (14) (1998).
377. M. Tolan, S.K. Sinha; unpublished.
378. M. Tolan, J. Wang, O. Bahr, J.Sokolov, M.H. Rafailovich, S.K. Sinha; unpublished.
379. H. Tostmann, E. DiMasi, P.S. Pershan, B.M. Ocko, O.G. Shpyrko, M. Deutsch; Phys. Rev. B (1998) in press.
380. S. Toxvaerd, J. Chem. Phys. **74**, 1998 (1981)
381. M.S. Turner and J.-F. Joanny; Macromolecules **25**, 6681 (1992).
382. M.S. Turner, M. Maaloum, D. Ausseré, J.-F. Joanny, M. Kunz; J. de Physique II **4**, 689 (1994).
383. J.H. van Zanten, W.E. Wallace, W.L. Wu; Phys. Rev. E **53**, R2053 (1996).
384. I.A. Vartanyants, J.A. Pitney, J.L. Libbert, I.K. Robinson; Phys. Rev. B **55**, 13193 (1997).
385. B. Vidal, P. Vincent; Applied Optics **23**, 1794 (1984).
386. G.H. Vineyard; Phys. Rev. B **26**, 4146 (1982).
387. J. Visser; Advances in Colloid Interface Science **15**, 157 (1981).
388. E. Vlieg, S.A. De Vries, J. Alvarez, S. Ferrer; J. Synchrotron Rad. **4**, 210 (1997).
389. M. Vossen, F. Forstmann; J. Chem. Phys. **101**, 2379 (1994).
390. W.E. Wallace, J.H van Zanten, W.L. Wu; Phys. Rev. E **52**, R3329 (1995).
391. W. Weber; PhD thesis, RWTH Aachen (1992).
392. W. Weber, B. Lengeler; Phys. Rev. B **46**, 7953 (1992).
393. A.H. Weiss, M. Deutsch, A. Braslau, B.M. Ocko, P.S. Pershan; Rev. Sci. Instrum. **57** (1986).
394. A.J.C. Wilson (ed.), *International Tables for Crystallography*, Kluwer Academic, Dordrecht, Boston, London (1992).
395. C.P. Wong (ed.); *Polymers for Electronic and Photonic Applications*, Academic Press, Boston (1993).
396. G.C.L. Wong, W.H. de Jeu, H. Shao, K.S. Liang, R. Zentel; Nature **389**, 567 (1997).
397. E.S. Wu, W.W. Wepp; Phys. Rev. A **8**, 2065 (1973).
398. W.-L. Wu, J.H. van Zanten, W.J. Orts; ISHM Proceedings, p. 507 (1994).
399. W.-L. Wu, W.E. Wallace, J. van Zanten; Mater. Res. Soc. Symp. Proc. **381**, 147 (1995).
400. X.Z. Wu, E.B. Sirota, S.K. Sinha, B.M. Ocko, M. Deutsch; Phys. Rev. Lett. **70**, 958 (1993).
401. X.Z. Wu, B.M. Ocko, E.B. Sirota, S.K. Sinha, M. Deutsch; Physica A **200**, 751 (1993).
402. X.Z. Wu, B.M. Ocko, E.B. Sirota, S.K. Sinha, M. Deutsch, H.B. Cao, M.W. Kim; Science **261**, 1018 (1993).
403. T.K. Xia, J. Ouyang, M.W. Ribarsky, U. Landman; Phys. Rev. Lett. **69**, 1967 (1992).
404. L. Xie, G.B. DeMaggio, W.E. Frieze, J. DeVries, D.W. Gidley, H.A. Hristov, A.F. Yee; Phys. Rev. Lett. **74**, 4947 (1995).
405. R. Xie, A. Karim, J.F. Douglas, C.C. Han, R.A. Weiss; Phys. Rev. Lett. (1998) in press.
406. A.M. Yaglom; *Correlation Theory of Stationary and Related Random Functions*, Springer Series in Statistics Vol. 1, Spinger, Berlin, Heidelberg (1987).
407. Y. Yoneda; Phys. Rev. **131**, 2010 (1963).
408. T. Young; Phil. Trans. R. Soc. Lond. **35**, 95 (1805).
409. H. Zabel; Physica B **198**, 156 (1994).
410. P. Zaumseil, U. Winter; phys. stat. sol. (a) **70**, 497 (1982).

411. W. Zhao, X. Zhao, M. Rafailovich, J. Sokolov, T. Mansfield, R.S. Stein, R.C. Composto, E.J. Kramer, R.A.L. Jones; Mater. Res. Soc. Symp. Proc. **171**, 337 (1990).
412. W. Zhao, X. Zhao, J. Sokolov, M.H Rafailovich, M.K. Sanyal, S.K. Sinha, B.H. Cao, M.W. Kim, B.B. Sauer; J. Chem. Phys. **97**, 8536 (1992).
413. W. Zhao, M.H Rafailovich, J. Sokolov, L.J. Fetters, R. Plano, M.K. Sanyal, S.K. Sinha, B.B. Sauer; Phys. Rev. Lett. **70**, 1453 (1993).
414. H. Zhao, A. Penninckx-Sans, L.-T. Lee, D. Beysens, G. Jannink; Phys. Rev. Lett. **75**, 1977 (1995).
415. Y.-P. Zhao, G.-C. Wang, T.-M. Lu; Phys. Rev. B **55**, 13938 (1997).
416. J. Zhu, A. Eisenberg, R.B. Lennox; J. Am. Chem. Soc. **113**, 5583 (1991).
417. J. Zhu, R.B. Lennox, A. Eisenberg; J. Phys. Chem. **96**, 4727 (1992).

Index

Printing: Mercedesdruck, Berlin
Binding: Buchbinderei Lüderitz & Bauer, Berlin

Springer Tracts in Modern Physics

130 **Time-Resolved Light Scattering from Excitons**
By H. Stolz 1994. 87 figs. XI, 210 pages

131 **Ultrathin Metal Films**
Magnetic and Structural Properties
By M. Wuttig, not yet published

132 **Interaction of Hydrogen Isotopes with Transition-Metals and Intermetallic Compounds**
By B. M. Andreev, E. P. Magomedbekov, G.H. Sicking 1996. 72 figs. VIII, 163 pages

133 **Matter at High Densities in Astrophysics**
Compact Stars and the Equation of State
In Honor of Friedrich Hund's 100th Birthday
By H. Riffert, H. Müther, H. Herold, and H. Ruder 1996. 86 figs. XIV, 278 pages

134 **Fermi Surfaces of Low-Dimensional Organic Metals and Superconductors**
By J. Wosnitza 1996. 88 figs. VIII, 172 pages

135 **From Coherent Tunneling to Relaxation**
Dissipative Quantum Dynamics of Interacting Defects
By A. Würger 1996. 51 figs. VIII, 216 pages

136 **Optical Properties of Semiconductor Quantum Dots**
By U. Woggon 1997. 126 figs. VIII, 252 pages

137 **The Mott Metal-Insulator Transition**
Models and Methods
By F. Gebhard 1997. 38 figs. XVI, 322 pages

138 **The Partonic Structure of the Photon**
Photoproduction at the Lepton-Proton Collider HERA
By M. Erdmann 1997. 54 figs. X, 118 pages

139 **Aharonov-Bohm and other Cyclic Phenomena**
By J. Hamilton 1997. 34 figs. X, 186 pages

140 **Exclusive Production of Neutral Vector Mesons at the Electron-Proton Collider HERA**
By J. A. Crittenden 1997. 34 figs. VIII, 108 pages

141 **Disordered Alloys**
Diffusive Scattering and Monte Carlo Simulations
By W. Schweika 1998. 48 figs. X, 126 pages

142 **Phonon Raman Scattering in Semiconductors, Quantum Wells and Superlattices**
Basic Results and Applications
By T. Ruf 1998. 143 figs. VIII, 252 pages

143 **Femtosecond Real-Time Spectroscopy of Small Molecules and Clusters**
By E. Schreiber 1998. 131 figs. XII, 212 pages

144 **New Aspects of Electromagnetic and Acoustic Wave Diffusion**
By POAN Research Group 1998. 31 figs. IX, 117 pages

145 **Handbook of Feynman Path Integrals**
By C. Grosche and F. Steiner. 1998. X, 449 pages

146 **Low-Energy Ion Irradiation of Solid Surfaces**
By H. Gnaser. 1999. 93 figs. VIII, 293 pages

147 **Dispersion, Complex Analysis and Optical Spectroscopy**
By K.-E. Peiponen, E.M. Vartiainen, and T. Asakura. 1999. 46 figs. VIII, 130 pages

148 **X-Ray Scattering from Soft-Matter Thin Films**
Materials Science and Basic Research
By M. Tolan. 1999. 98 figs. IX, 197 pages